THE
DANCE
of
THE
MOON

THE DANCE OF THE MOON

MOON

Pari Spolter

Orb Publishing Company
Granada Hills, California

Orb Publishing Company
16911 San Fernando Mission #182
Granada Hills, CA 91344 USA

Cover photo: *NASA Photograph of Eclipses;*
http://eclipse.gsfc.nasa.gov/

Publisher's Cataloging in Publication Data:

Spolter, Pari.
    The dance of the moon / Pari Spolter.
    pages cm
    Includes bibliographical references and index.
    ISBN 978-0-9638107-8-6

    1. Lunar theory. 2. Moon--Orbit. 3. Gravitation.
I. Title.

QB391.S66 2015   523.3'3
                  QBI15-600002

Library of Congress Control Number: 2015900703

In loving memory of my late husband
Herbert Spolter, M.D.

Preface

In this book, you will not find a long series of advanced mathematical equations. There are no theory or assumptions. Instead, you will find many easy to understand graphs based on the latest ephemerides of unprecedented accuracy.

The search to understand and explain the complicated motions of the Moon, our closest celestial neighbor, has been the subject of extensive work by numerous luminaries of the past. This has been a very interesting problem to me. I have enjoyed working on it.

I would not have been able to complete the work presented in this book in one lifetime, if not for the advances in technology. The invention of the atomic clock, the advent of lunar laser ranging technique, and the availability of computers, of small handheld scientific programmable calculators, of software for plotting and statistical calculations, have made the work presented in this book possible.

I wish to thank the people who have sent letters and emails to me for their interest in my first book *Gravitational Force of the Sun,* and my publications:

1. *New Concepts in Gravitation,* published in Physics Essays, Volume 18, 2005, pages 37–49.
2. *Problems with the Gravitational Constant,* published in Infinite Energy, Issue 59, 2005, page 39.
3. *Kepler's Second Law and Conservation of Angular Momentum,* published in Physics Essays, Volume 24, 2011, pages 260–266.

Financial support from my deceased husband has made the work presented in this book possible.

Granada Hills, California   Pari Spolter
January 2015

The Moonlight Sonata
Ludwig van Beethoven

Contents

Preface   vii

1. Introduction   1
The Puzzling Motion and efforts to explain it   1
The Peculiarity of the Orbit of the Moon   3
Calculations   3

2. Moon's Periods   13
Variation of Periods   13
Coincidence of Periods   17

3. Saros   44
Variation of Period   115

4. Perturbation by the Sun   119
Angular Momentum   136

5. Regression of the Nodes   143

6. Advance of the Perigee   156

7. The Obliquity of the Ecliptic   171

Index   189

CHAPTER 1

Introduction

The puzzling motion of the Moon and efforts to explain it

$\mathcal{O}_{ur}$closest celestial neighbor the Moon is visible both during the night and during the day. Its complex motions have attracted the attention of sky watchers for thousands of year. The history of the attempts to explain the Moon motion's vagaries is skillfully presented in an article by Martin Charles Gutzwiller (1925–2014) published in 1998.[1]

The motion of the moon problem commanded the attention of Isaac Newton (1642–1727) for many years. He remarked that the theory of the moon made his head ache and kept him awake so often that he would think of it no more.[2,3]

Newton's *Theory of the Moon's Motion* was published in 1702.[4] Ninety three years after Johannes Kepler (1571–1630) had banished epicycles from the heavens with a new physical astronomy; Newton had regressed to a kinematic approach, with the old epicycles and deferents still there.[5] A revised and much expanded version of the *Theory of the Moon's Motion* was published as the new Scholium to Proposition 35, Book 3 in the second (1713) and third (1726) editions of the *Principia*.[6] Derek Thomas Whiteside (1932–2008), professor of History of Mathematics and Exact Sciences at Cambridge University, the foremost authority on the work of Isaac Newton and editor of "The Mathematical Papers of Isaac Newton," in his article "Newton's Lunar Theory: From High Hope to Disenchantment" concludes:[7]

Pity those–notably Halley–who in the early decades of the eighteenth century tried to found solidly accurate tables of the moon's motion upon such a flimsy, rickety basis.

The difficulty of the problem was a stimulus to ingenuity of the ablest mathematicians of the eighteenth–century. Leonard Euler (1707–1783), a mathematical genius, wrote in the Preface of his last great work on lunar theory, in 1772:[8]

As often as I have tried these forty years to derive the theory and motion of the moon from the principles of gravitation, there always arose so many difficulties that I am compelled to break off my work and latest researches.

François–Félix Tisserand (1845–1896) in the third of his classic four–volume *Traité de Mécanique Céleste (Treatise on Celestial Mechanics): "Exposé de l'ensemble des Théories Relatives au Mouvement de la Lune" (The Display of all the Theories Relating to the Motion of the Moon)*[9] acknowledges some fundamental difficulties and hopes for a major discovery without knowing where it could come from.[10] Gutzwiller states:[11]

At the end of the 1990s it has to be admitted that Tisserand's diagnosis is still valid..

Archie Edminston Roy (1924–2012), Professor of Astronomy at the University of Glasgow, and author of the book *Orbital Motion* states:[12]

The construction of a complete lunar theory which not only includes the effects of Earth, Sun, planets and the figures of Earth and Moon but can also be compared with observations is one of the most difficult in astronomy.

In 1994 Martin Gutzwiller presented a paper at the American Institute of Physics Conference in Williamsburg, Virginia entitled: "Moon–Earth–Sun: The Oldest, Best Known, but Least Understood Three Body Problem."[13] In an article published in April 2013 of the *Notices of the American Mathematical Society* Florin Diacu, professor of Mathematics at the University of Victoria, Canada indicates[14]

Understanding the moon's orbit around Earth is a difficult mathematical problem...[The] moon's orbit is not fully explained today.

The Peculiarity of the Orbit of the Moon

The motion of the Moon around the Earth is different from the motion of a planet around the Sun in the following ways:

a. The orbit of the Moon is not in the equatorial plane of the Earth; thus in addition to the gravitational force, angular momentum has also to be considered.

b. The attraction between the Earth and the Moon is mutual. The attraction of the Earth by the Moon cannot be ignored. The Earth's ocean tides were first explained by Johannes Kepler (1571-1630) to be due to the Moon's gravitational force.[15] The center of gravity between the Earth and the Sun is at 6.456×10^{-4} of the Sun's radius. However, the center of gravity between the Moon and the Earth is at 0.73 of the Earth's radius.

c. The obliquity of the ecliptic affects the motion of the Moon. See chapter 7.

Calculations

The distinguished English physicist Louis Essen (1908-1997) invented the Atomic Clock in the 1950s at the National Physical Laboratory at Teddington in Middlesex, available at http://www.npl.co.uk/people/louis-essen.

Lasers became available in the early 1960s. The term **laser** is an acronym for light amplification by stimulated emission of radiation. The Apollo 11 package containing 100 solid fused silica corner reflectors, Figure 1.1, was placed on the lunar surface by the US astronauts Neil Armstrong and Edwin Aldrin on 21 July 1969. Successful range measurements were obtained by the Lick Observatory of the University of California; McDonald Observatory of the University of Texas; The Air

Force Cambridge Research Laboratories Lunar Ranging Observatory in Arizona; the Pic du Midi Observatory in France; and the Tokyo Astronomical Observatory in Japan.[16]

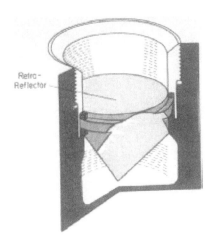

Retro-
Reflector

Figure 1.1

Cutaway view of a cube corner retroreflector mounted in the Apollo 11 package. Two cube corner reflectors are made by cutting a nearly perfect cube of fused silica in half across a body diagonal, and then polishing the resulting new faces flat. Light entering the front face is reflected from all three mutually perpendicular rear faces of the corner, and then goes back in the same direction it came from.

The Apollo 14 astronauts placed an array of 100 corner reflectors on the Moon on 5 February 1971, and the Apollo 15 astronauts placed 300 reflectors on the Moon on 31 July 1971. See Figure 1.2 A and B.[17]

Figure 1.2
(A) Photograph of a retroreflector array of Apollo 14 on the lunar surface.
(B) Geographical distribution of the retroreflector arrays on the lunar
surface. The labels A-11, A-14, and A-15 denote the Apollo 11, 14, and 15
sites, respectively, whereas L-1 and L-2 indicate the Lunakhod 1 and 2
locations (no returns are available from Lunakhod 1).

Lunar Laser Ranging (LLR) is based on measurements of the round-trip travel time of the laser pulses sent from Earth to the retro reflectors on the Moon. The measurements of about 2.4 to 2.7 seconds are made with a Cesium Atomic Clock. The Earth-Moon distance is calculated from the known value of the speed of light: $c = 299,792,500$ m s^{-1}.

The 2.72 meter McDonald telescope made LLR observations from 1970 until 1985 using ruby laser pulses and achieved 15 cm range precision.[18] Ruby is crystalline Aluminum Oxide, Al_2O_3, a small number of its Al^{+++} ions are replaced by Cr^{+++} ions, which produces red light with wavelengths of 693 to 705 $nanometers$. The 0.76 m telescope at McDonald Laser Ranging Station (MLRS) in Saddle and Mt Fowlkes in Texas made LLR observations using Nd:YAG laser until 2007. Neodymium ions in crystals of Yttrium Aluminum Garnet produce laser wavelengths of 1064 $nanometer$ (infrared) and 532 nm (green).

The French station Centre d'Étude et de Recherche en Géodynamique et Astronomie (CERGA) at the Observatoire de la Côte d'Azur (OCA) near Grass was built in 1984 with a 1.54 m telescope. The precision at the beginning was in the 15 cm range. OCA replaced its ruby laser by a Nd:YAG in 1987 and has reached a range precision of 2.5 cm.[19,20,21]

The ingenious construction of the Apache Point Observatory Lunar Laser-ranging Operation (APOLLO acronym) in southern New Mexico is described by Murphy et al.[22] The 3.5 m telescope uses Nd: YAG laser. APOLLO was built in April 2006. After several improvements, it has achieved a range precision of 1.1 $millimeter$ for the Earth-Moon distance since September 2007.[23]

Since 1981, The Astronomical Almanac[24] includes the Lunar Polynomial Tables. These data allow calculations of the maximums and the minimums at any instant of a day. Figure 1.3 is the geocentric equatorial coordinate system.

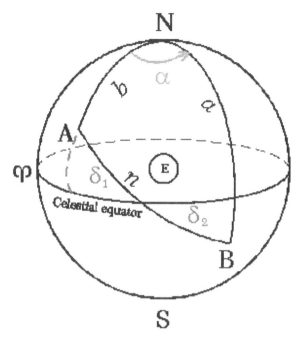

Figure 1.3
The geocentric equatorial coordinate system

The law of cosine for a side n of the spherical triangle ANB is:[25,26]

$$\cos n = \cos a \, \cos b + \sin a \, \sin b \cos \alpha \qquad (1.1)$$

Sides a and b are $90° \pm$ declinations δ_2 and δ_1.
α is the Right Ascension $\alpha_2 - \alpha_1$. In one day, the Moon moves from α_1 to α_2.
n is the arc of the spherical triangle ANB measured in degrees.
$n \times \pi / 180 = \theta$ in radians.
$\theta \times r = d$
r is the medium of the Earth – Moon distance and d is the daily distance traveled by the Moon.
The daily velocity of the Moon is d /86400 in $m \, s^{-1}$.

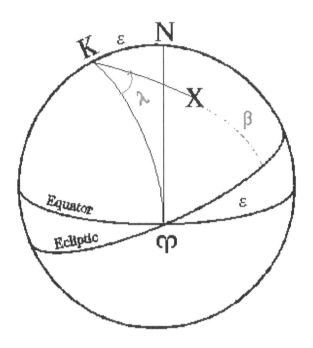

Figure 1.4
The geocentric ecliptic coordinate system. N is the north celestial pole, and K is the north pole of the ecliptic.

The transformation equations are:[27]

$$\sin \beta = \sin \delta \cos \varepsilon - \cos \delta \sin \varepsilon \sin \alpha$$

$$\cos \lambda = \frac{\cos \delta \cos \alpha}{\cos \beta} \qquad (1.2)$$

λ is the longitude, β is the latitude, and ε is the obliquity of the ecliptic.

References

1. Martin C. Gutzwiller, "Moon-Earth-Sun: The oldest three-body problem," *Reviews of Modern Physics,* **70** (1998), pp. 589-639.

2. Sir David Brewater, *Memoirs of the Life, Writings, and Discoveries of Sir Isaac Newton* (New York: Johnson Reprint Corporation, 1965), vol. 1, pp. 157-158.

3. Richard S. Westfall, *Never at Rest – A Biography of Isaac Newton* (Cambridge University Press, 1980), p. 544.

4. Nicholas Kollerstrom, *Newton's Forgotten Lunar Theory* (Santa Fe, New Mexico: Green Lion Press, 2000).

5. Ibid., p. 32 and p. 179.

6. Craig B. Waff, "Newton and the Motion of the Moon–An Essay Review," *Centaurus,* **21** (1977), pp. 64-75 at p.71.

7. D. T. Whiteside, "Newton's Lunar Theory: From High Hope to Disenchantment," *Vista in Astronomy,* **19** (1976), pp. 317-328 at p. 324.

8. Antonie Pannekoek, *A History of Astronomy* (New York: Dover Publications, Inc. 1989), p. 299.

9. François–Félix Tisserand, *Traité de Mécanique Célest,* Tome III: *Exposé de l'ensemble des Théories Relatives au Mouvement de la Lune* (Paris: Gauthier–Villars et fils, 1894).

10. Martin C. Gutzwiller, op. cit., p. 624.

11. Ibid., p. 625.

12. Archie Edminston Roy, *Orbital Motion* (Bristol and Philadelphia: Institute of Physics Publishing, 1988), Third edition, p. 281.

13. Martin C. Gutzwiller, "Moon–Eart–Sun: The Oldest, Best Known, but Least Understood Three body–Problem,"American Institute of Physics Conference Proceeding 334, May 26–31, 1994. Few–Body Problems in Physics, Williamsburg, VA. Editor: Franz Gross.

14. Florin Diacu, "Mathematical Methods in the Study of Historical Chronology," *Notices of the American Mathematical Society* **60** (2013), pp.441–449, at page 443.

15. William H. Donahue, *Johannes Kepler NEW ASTRONOMY* (Cambridge University Press, 1992), pp. 56 and 57.

16. P. L. Bender, D. G. Currie, R. H. Dicke, D. H. Eckhardt, J. E. Faller, W. M. Kaula, J. D. Mulholland, H. H. Pkotkin, S. K. Poultney, E. C. Solverberg, D. T. Wilkinson, J. G. Williams, C. O. Alley, "The Lunar Laser Ranging Experiment," *Science* **182** (1973), pp. 229–238. Copyright © at page 230 Reprinted by permission of *Science* Magazine.

17. J. O. Dickey, P. L. Bender, J. E. Faller, X. X. Newhall, R. L. Ricklefs, J. G. Ries, P. J. Shelus, C. Veillet, A. L. Whipple, J. R. Wiant, J. G. Williams, C. F. Yoder, "Lunar Laser Ranging: A Continuing Legacy of the Apollo Program," *Science* **265** (1994), pp. 482–490. Copyright © at page 483 Reprinted by permission of *Science* Magazine.

18. C. O. Alley, R. F. Chang, D. G. Currie, S. K. Poultney, P. L. Bender, R. H. Dicke, D. T. Wilkinson, J. E. Faller, W. M. Kaula, G. J. F. MacDonald, J. D. Mulholland, H. H. Plotkin, W. Carrion, E. J. Wampler, "Laser Ranging Retro–Reflector: Continuing Measurements and Expected

Results," *Science* **167** (1970), pp. 458–460.

19 E. Samain, J. F. Mangin, C. Veillet, J. M. Torre, P. Fridelance, J. E. Chabaudie, D. Féraudy, M. Glentzlin, J. Pham Van, M. Furia, A. Journet, and G. Vigouroux, "Millimetric Lunar Laser Ranging at OCA (Observatoire de la Côte d'Azur)," *Astronomy & Astrophysics Supplement* **130** (1998), pp. 235–244.

20. J. Chapront, M. Chapront–Touzé, and G. Francou, "A new determination of lunar orbital parameters, precession constant and tidal acceleration from LLR measurements," *Astronomy & Astrophysics* **387** (2002), pp. 700–709.

21. Jean Chapront, Michelle Chapront–Touzé, and Gerard Francou, Patrick Bidard, Jean–François Mangin, Dominique Feraudy, Maurice Furia, Alain Journet, Jean–Marie Torre, Gérard Vigouroux,
cddis.gsfc.nasa.gov/lw12/docs/chapront_et_al_JC_JFM.pdf

22. T. W. Murphy, Jr., E. G. Adelberger, J. B. R. Battat, L. N. Carey, C. D. Hoyle, P. LeBlanc, E. L. Michelsen, K. Nordtvedt, A. E. Orin, J. D. Strasburg, C. W. Stubbs, H. E. Swanson, and E. Williams, "The Apache Point Observatory Lunar Laser–ranging Operation: Instrument Description and First Detections," *Publications of the Astronomical Society of the Pacific* **120** (2008), pp.20–37.

23. J. B. R. Battat, T. W. Murphy, Jr., E. G. Adelberger, B. Gillespie, C. D. Hoyle, R. J. McMillan, E. L. Michelsen, K. Nordtvedt, A. E. Orin, C. W. Stubbs, and H. E. Swanson, "The Apache Point Observatory Lunar Laser–ranging Operation (APOLLO): Two Years of Millimeter–Precision Measurements of the Earth–Moon Range,"*Publications of the Astronomical Society of the Pacific* **121** (2009), pp.29–40.

24. *The Astronomical Almanac* (Washington D.C.: U.S. Government Printing Office.)

25. Robin M. Green, *Spherical astronomy* (Cambridge University Press, 1985), p. 9 and p.12.

26. Kaj L. Nielsen, *Modern Trigonometry* (Barnes & Noble Books, 1966), p. 180.

27 Robin M. Green, *op. cit.,* p. 34.

CHAPTER 2

Moon's Periods

Variation of Periods

𝒯he *Astronomical Almanac*[1] gives the lengths of the **mean** months, as derived from the mean orbital elements as:

		Days
synodic month	(new moon to new moon)	29.530589
tropical month	(equinox to equinox)	27.321582
sidereal month	(fixed star to fixed star)	27.321662
anomalistic month	(perigee to perigee)	27.554550
draconic momth	(node to node)	27.212221

The orbital periods vary from these mean values. Figure 2.1 shows periodical variations of New Moon to New Moon and of Full Moon to Full Moon for 30 years from 1981 to 2010; 370 cycles.

For New Moon to New Moon at this interval, the Minimum value is 29.288889 days, and the Maximum value is 29.826389 days. The arithmetic mean is 29.528155 days ± 0.00767 standard error.

For Full Moon to Full Moon at this interval, the Minimum value is 29.275694 days, and the Maximum value is 29.809722 days. The arithmetic mean is 29.532486 days ± 0.00719 standard error.

Figure 2.2 shows periodical variations of First Quarter to First Quarter and of Last Quarter to Last Quarter for 30 years from 1981 to 2010; 370 cycles.

For First Quarter to First Quarter at this interval, the Minimum value is 29.179167 days, and the Maximum value is 29.918056 days. The arithmetic mean is 29.529688 days ± 0.01081 standard error.

For Last Quarter to Last Quarter at this interval, the Minimum value is 29.181944 days, and the Maximum value is 29.925694 days. The arithmetic mean is 29.529755 days ± 0.01082 standard error.

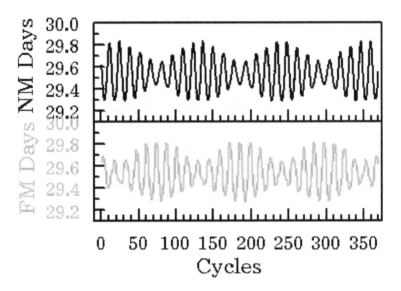

Figure 2.1 New Moon to New Moon and Full Moon to Full Moon Period.

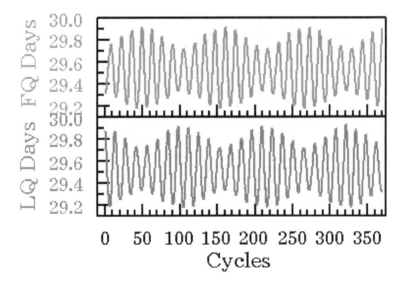

Figure 2.2 First Quarter to First Quarter and Last Quarter to Last Quarter Periods.

The periodical variations of Sidereal month are shown in Figure 2.3 for 28 years from 1981 to 2008; 373 Cycles. The Minimum value at this interval is 27.213933 days, and the Maximum value is 27.468415 days. The arithmetic mean is 27.32318 days ± 0.00355 standard error.

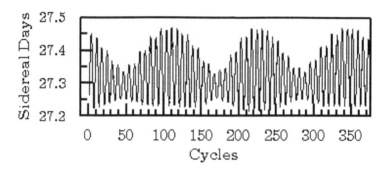

Figure 2.3 Variations of Periods of Sidereal Months.

The periodical variations of anomalistic month, Perigee to Perigee, and Apogee to Apogee are shown in Figure 2.4 for 28 years from 1981 to 2008; 370 cycles.

For Perigee to Perigee at this interval the Minimum value is 24.65095 days, and the Maximum value is 28.55735 days. The arithmetic mean is 27.550320 days ± 0.058650 standard error.

For Apogee to Apogee at this interval the Minimum value is 26.98475 days, and the Maximum value is 27.8958 days. The arithmetic mean is 27.553750 ± 0.013951 standard error.

The variation of the period of Perigee is much greater than the variation of the period of Apogee. This will be discussed in chapter 6 on Perigee and Apogee.

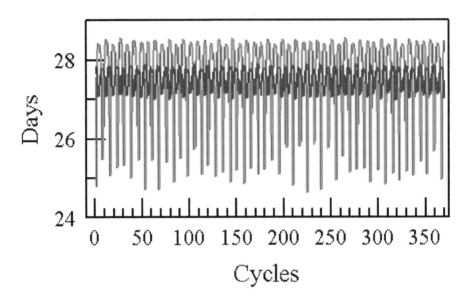

Figure 2.4
Variarion of Perigee to Perigee and of Apogee to Apogee Periods.

The periodical variations of the draconic month (node to node) are shown in Figure 2.5 for 28 years from 1981 to 2008; 375 cycles.

For Descending Node of the latitude to Descending Node at this interval the Minimum value is 27.013990 days, and the Maximum value is 27.483467 days. The arithmetic mean is 27.210015 days ± 0.006382 standard error.

For Ascending Node of the latitude to Ascending Node at this interval the Minimum value is 27.006422 days, and the Maximum value is 27.469316 days. The arithmetic mean is 27.213874 days ± 0.006642 standard error.

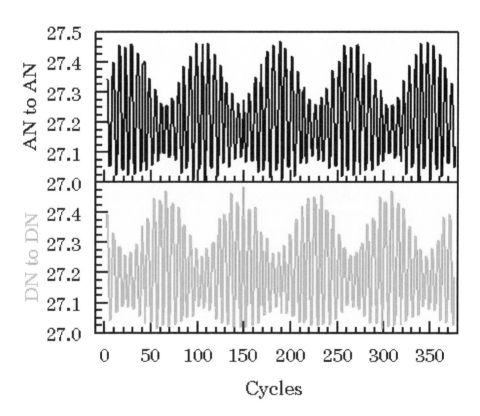

Figure 2.5
Variation of Descending Node of the Latitude to Descending Node and of
Ascending Node to Ascending Node Periods.

Coincidence of Periods

The mean synodic month of $29.530589 \times 14 = 413.428246$ days.
The mean anomalistic month of $27.554550 \times 15 = 413.31825$ days. In
Figure 2.6 cycles 1 to 24 are shown. Cycle 1 starts at Perigee of May
4.20325, and at New Moon of May 4.17986 in 1981. Cycle 24 ends at
Perigee of July 1.8946, and at New Moon of July 3.0965278 in 2008.

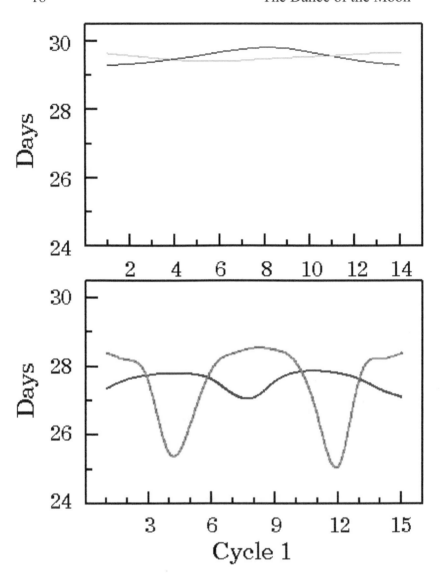

Figure 2.6
New Moon to New Moon, Full Moon to Full Moon, Perigee to Perigee, and Apogee to Apogee Periods for Cycles 1 through 24.

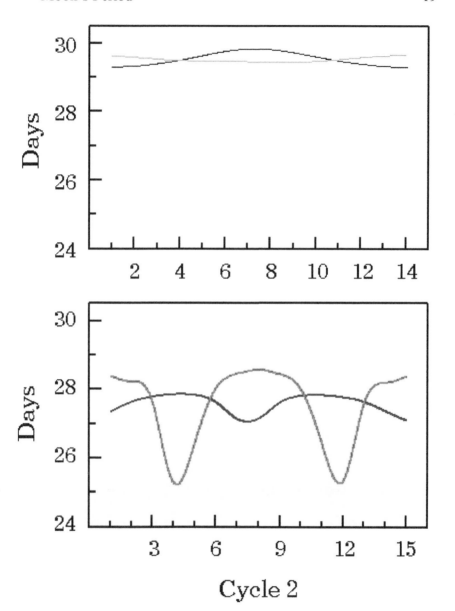

Cycle 2

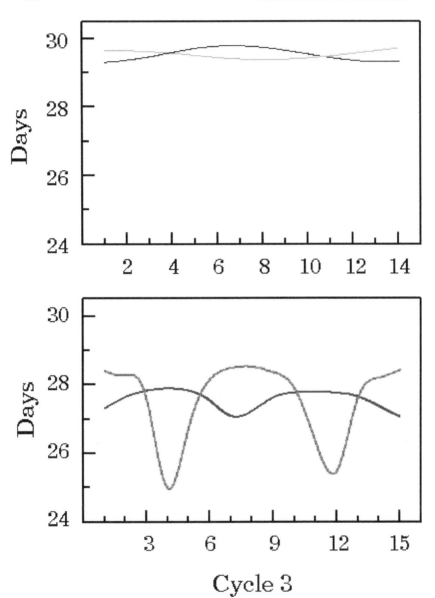

Cycle 3

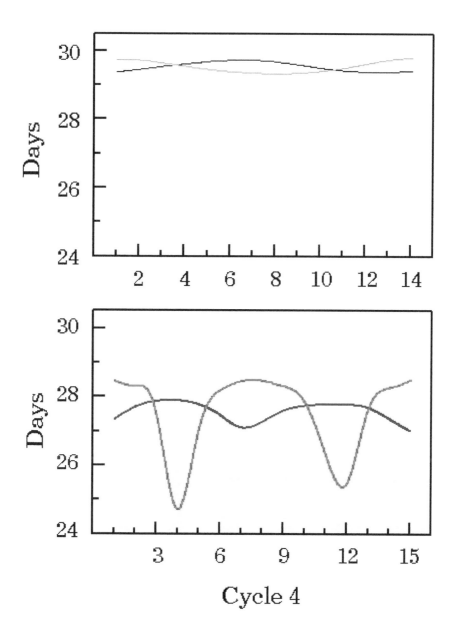

Cycle 4

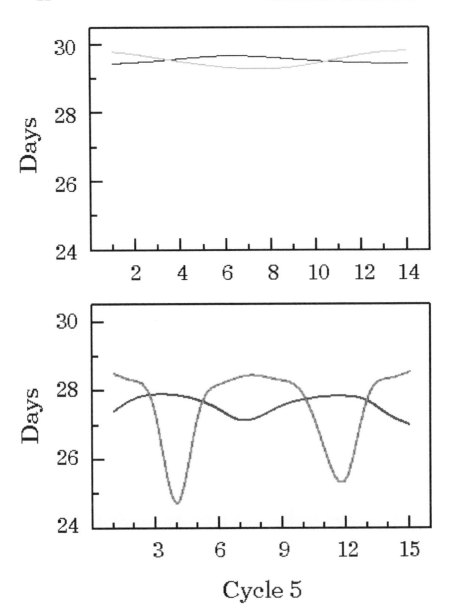

Cycle 5

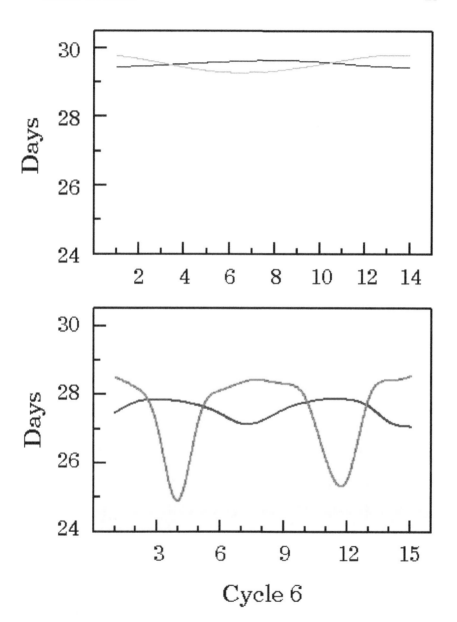

Cycle 6

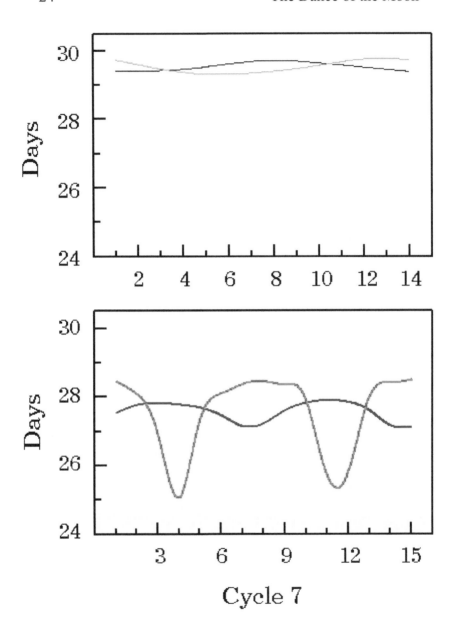

Cycle 7

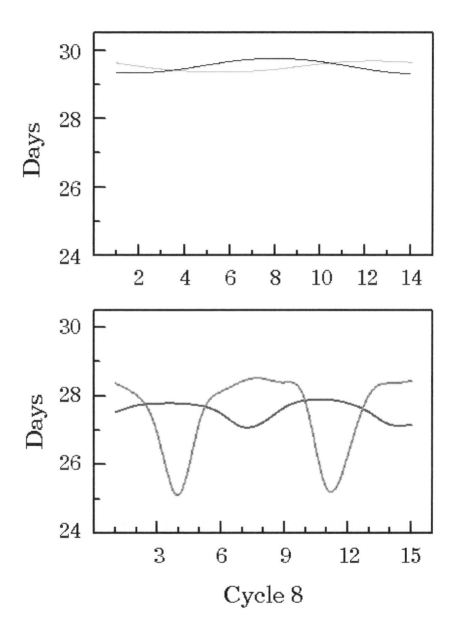

Cycle 8

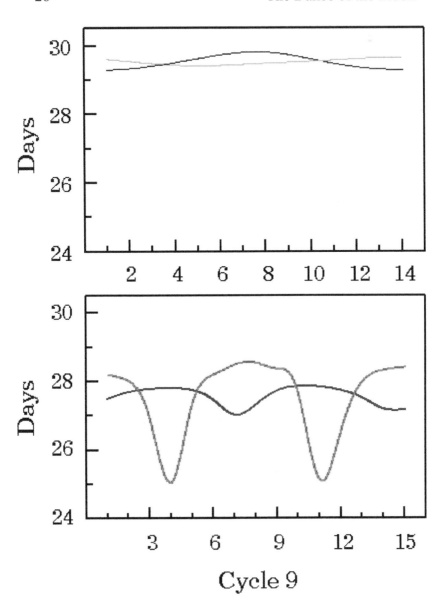

Cycle 9

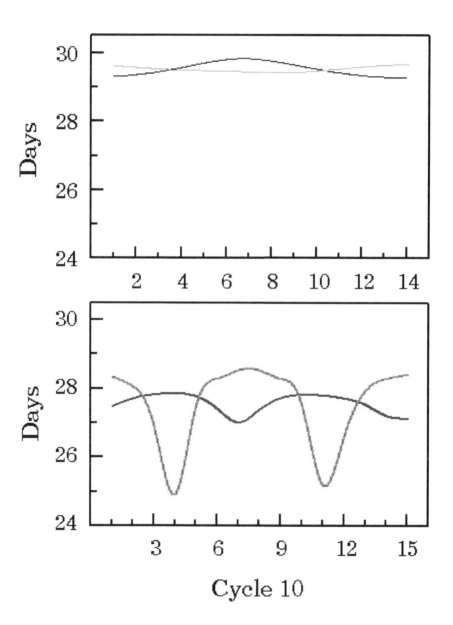

Cycle 10

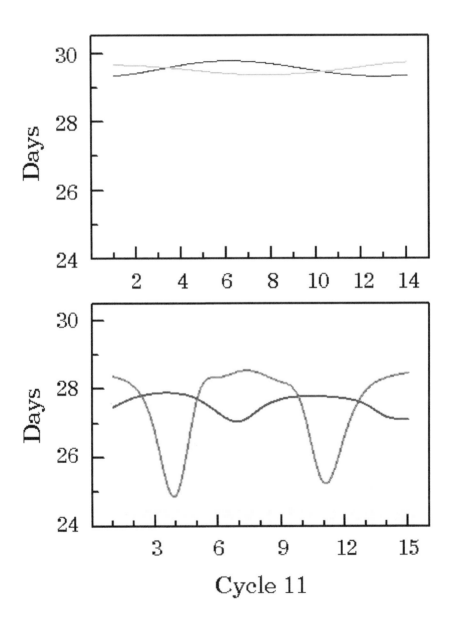

Cycle 11

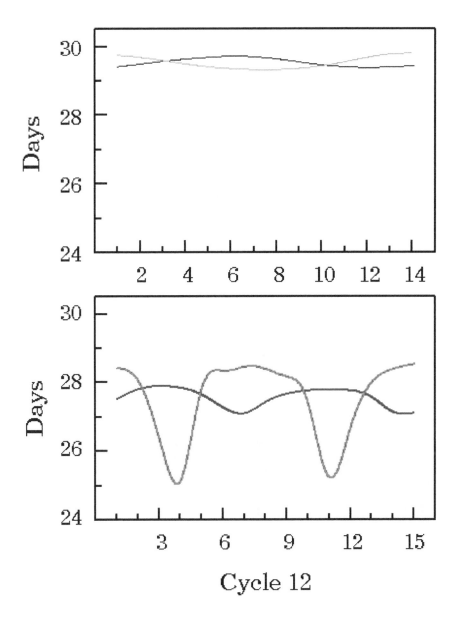

Cycle 12

The Dance of the Moon

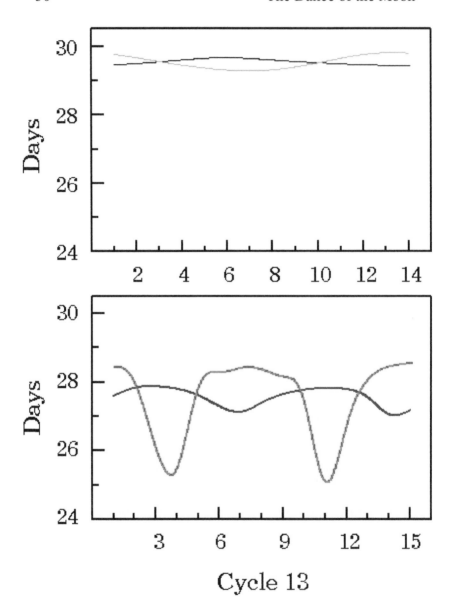

Cycle 13

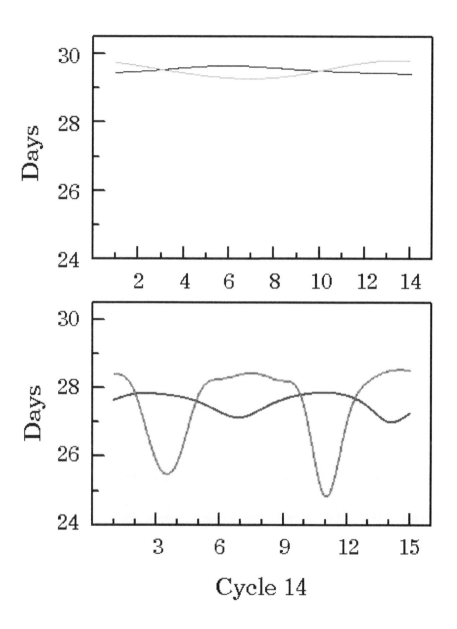

Cycle 14

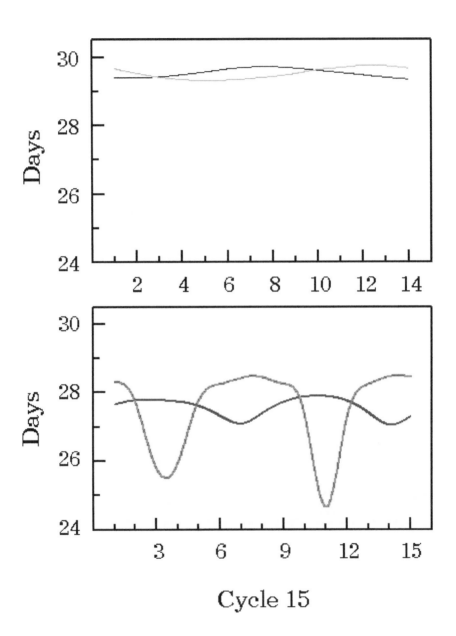

Cycle 15

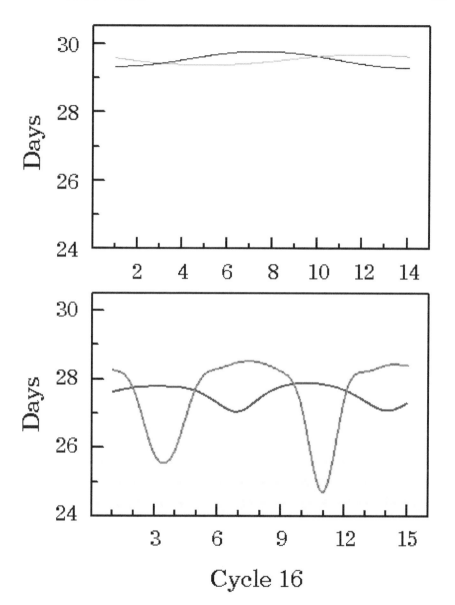

Cycle 16

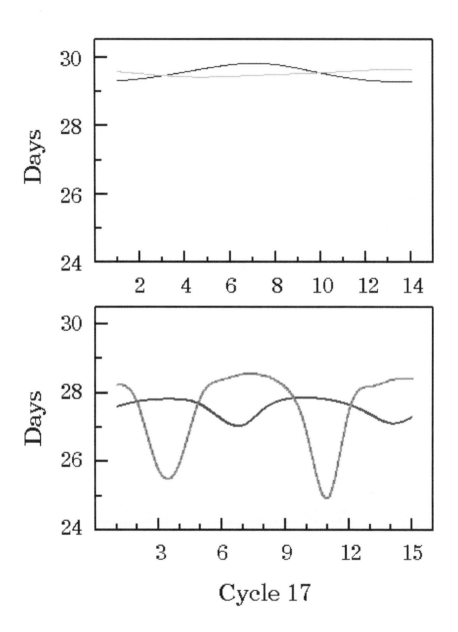

Cycle 17

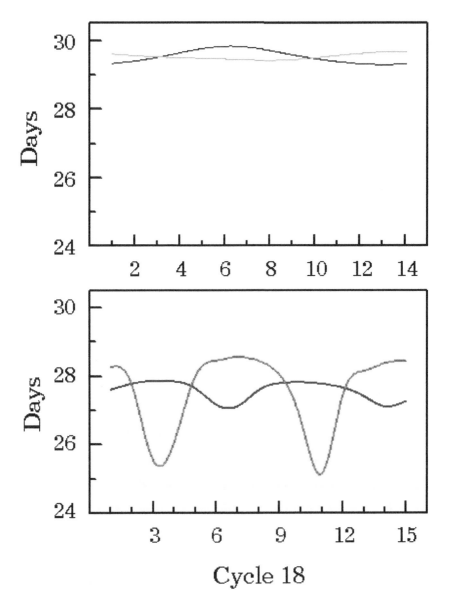

Cycle 18

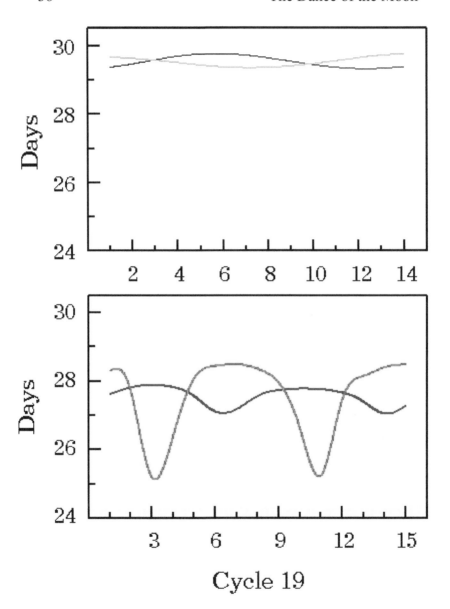

Cycle 19

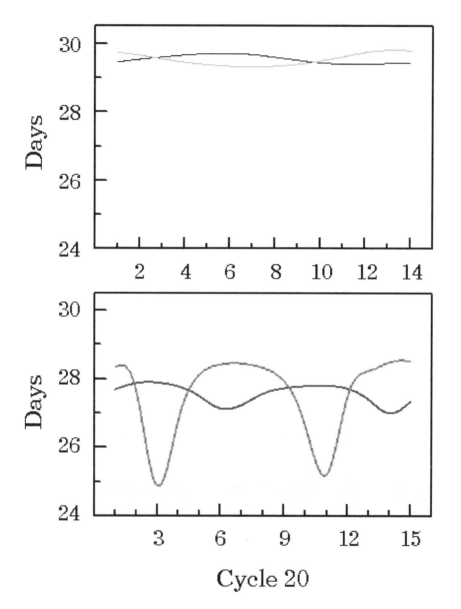

Cycle 20

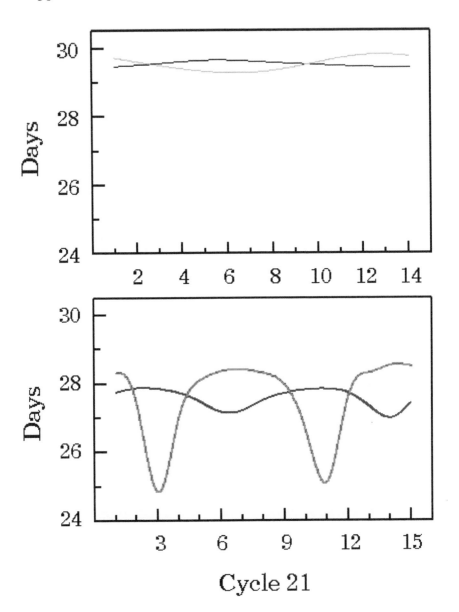

Cycle 21

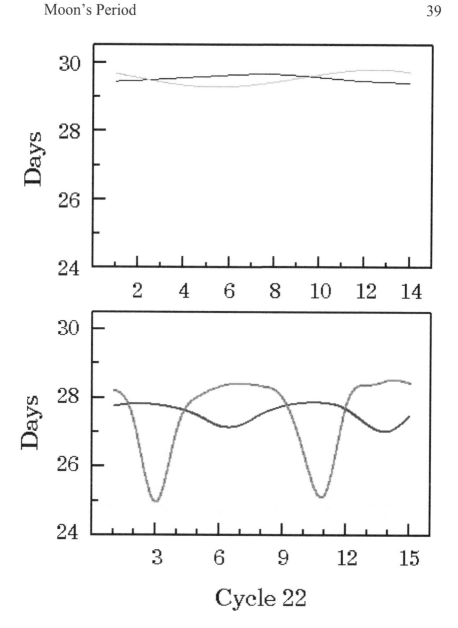

Cycle 22

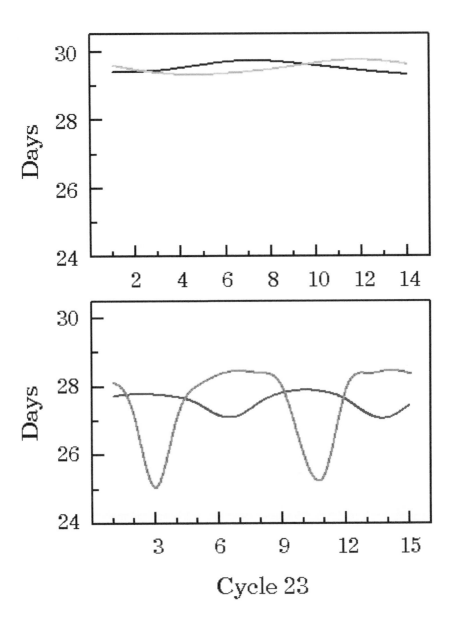

Cycle 23

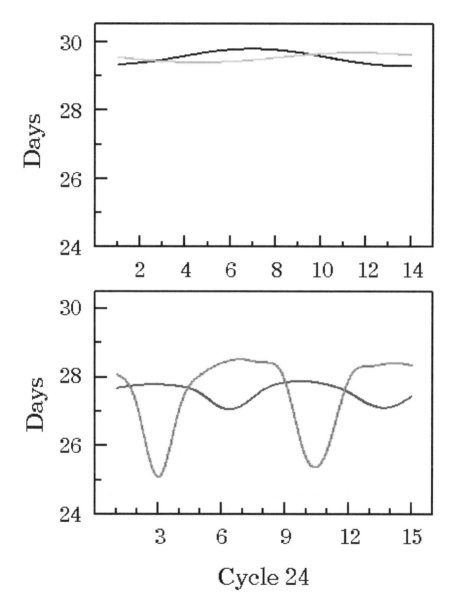

Cycle 24

Table 2.1

15 Perigees and 14 New Moons

Cycles	15 Perigees, days	14 New Moons, days
1	413.2928	413.309028
2	413.32535	413.318055
3	413.36085	413.39792
4	413.4535	413.531944
5	413.43655	413.53472
6	413.38615	413.5125
7	413.33775	413.4056
8	413.3111	413.34306
9	413.2935	413.30486
10	413.315	413.3167
11	413.3397	413.37083
12	413.43785	413.5125
13	413.4011	413.52361
14	413.33865	413.48056
15	413.29975	413.35417
16	413.2555	413.2861
17	413.32575	413.32986
18	413.37705	413.3778

19	413.40995	413.50764
20	413.4404	413.36319
21	413.34325	413.50327
22	413.28735	413.44028
23	413.27395	413.31181
24	413.2609	413.29167

More coincidence of Moon's months will be presented in next chapter on Saros.

Reference

1. *The Astronomical Almanac* for the years 1981 to 2015 (Washington, D.C. US Government Printing Office) p. D2.

CHAPTER 3

SAROS

𝒯ℎₑ most famous of the eclipse rhythms, the Saros, is a set of approximate commensurabilities between the **mean** synodic, draconic, and anomalistic lunar months:

223 Synodic months of 29.530589 days   =   6585.321347 days

242 Draconic months of 27.212221 days   =   6585.357482 days

239 Anomalistic months of 27.55455 days = 6585.537450 days

The interval is about 18 years and 11⅓ days (10⅓ days when five leap years are included). This period was discovered by the Chaldeans (ancient Babylonian astronomers) in the fourth-century BC. It was referred to as the 18-year cycle. Edmond Halley (1656-1742) named the cycle Saros in 1691. For the origin of this misnomer see Neugebauer.[1]

Lunar eclipses occur at Full Moons when the Earth shadow covers the Moon. The total phase of lunar eclipse can last as long as 1 hour 46 minutes,[2] and is visible from the entire night side of the Earth. When totally eclipsed, the Moon has a dull coppery red color, see Figure 3.1. Kepler explained in his *Epitome* the cause of it is sunlight refracted by the Earth's atmosphere.[3]

Solar eclipses occur at New Moons when the Moon comes directly between the Earth and the Sun, see Figure 3.2. By cosmic coincidence, on the average, the Sun is 389.1 times farther from Earth than the Moon and 400.5 times bigger than the Moon.[4] Because the Moon's umbra just barely reaches the Earth, the width of the solar eclipse path, within which a total eclipse can be seen, is very small. The totality of a solar eclipse can never exceed 7 minutes 31 seconds.[5]

The configuration of the Earth-Moon-Sun at any time, and not just at times of eclipses, is very nearly repeated after one Saros period.[6,7]

Figure 3.1

NASA photograph of total eclipse of the Moon Tuesday, April 15, 2014. Saros 122 (56/75).

Figure 3.2

Total Solar eclipse of August 11, 1999. Photograph from Wikipedia, the free encyclopedia. At the moment of totality, Sun's corona, the Baily beads and the red prominences are visible.

Eclipses migrate westward by approximately 120° after each Saros period. This results from the fact that the length of one Saros is an integer number of days *plus one third of a day*. After three Saroses the central line will again be near the same geographical longitude as the starting eclipse.[8] The Earth rotates exactly 6585.321 times around its own axis in 223 synodic months. If we wait for three Saros cycles, the Earth will have rotated almost exactly back to its previous orientation, and the eclipse will be seen again from the same place at the same time of the day.[9]

The 223 synodic months are not exactly equal to an integer number, 242, of draconic months. Consequently, after one Saros the New or Full Moon will not be at exactly the same distance from the ascending or descending node of the lunar orbit. After several repetitions, the New or Full Moon will be too far from the lunar node and no longer will an eclipse be possible. That Saros series will then have come to an end.

A Saros series starts with a small partial eclipse visible near one of the poles of the Earth, north or south. Solar eclipses occurring near the Moon's descending node are given *even* Saros series numbers. The first eclipse of each series starts at the southern pole of the Earth. The next eclipses of that series occur gradually closer to the node of the lunar orbit and successive members work their way from south to north. *Odd* numbers are used for solar eclipses occurring near the ascending node, and the first eclipse of each series starts at the north pole of the Earth; they work their way from north to south.[10,11]

A Saros series lasts between 1,226 and 1,550 years and comprises 69 to 87 eclipses. As one series end, another is born. On average, 42 series are running in parallel at a given time.[12]

The length of a Saros period is not a fixed value. The 18 years and 11⅓ days is an average.[13] Calculations of 67 Saros periods from 1981 to 2013 are presented in the next pages. The Saros series numbers and sequences are from the two books of Canons of Solar and Lunar Eclipses by Fred Espenak, nicknamed NASA's Mr. Eclipse.[14,15]

In the following tables α is the Right Ascension, and δ is the Declination in the geocentric equatorial coordinate system; λ is the Longitude, and β is the Latitude in the geocentric ecliptic coordinate system. See chapter 1, pages 7–9.

1981
Penumbral Eclipse of the Moon
January 20.3354726852

Saros Number 114 (57/71)

Full Moon: January 20.31875
Earth–Moon Distance: 379,676,334.5 m
$\alpha = 122.°3692096$ $\delta = + 19.°12901744$
$\lambda = 120.°3899457$ $\beta = - 0.°9604411106$

1999
Penumbral Eclipse of the Moon
January 31.6930807871

Saros Number 114 (58/71)

Full Moon: January 31.670833
Earth–Moon Distance: 378,596,559.6 m
$\alpha = 133.°7984701$ $\delta = + 16.°37397542$
$\lambda = 131.°616901$ $\beta = - 0.°9637143906$

Saros Period = 6585.3576081019 days

1981
Annular Eclipse of the Sun
February 4.9150209492

Saros Number 140 (27/71)

New Moon: February 4.926389
Earth–Moon Distance: 375,977,210.1 m
$\alpha = 318.°4664444 \qquad \delta = -16.°51717386$
$\lambda = 315.°8649454 \qquad \beta = -0.°4570451675$

1999
Annular Eclipse of the Sun
February 16.2641099421

Saros Number 140 (28/71)

New Moon: February 16.2770833
Earth–Moon Distance: 377,102,906.9 m
$\alpha = 329.°3182812 \qquad \delta = -12.°94406236$
$\lambda = 326.°9490921 \qquad \beta = -0.°4411950792$

Saros Period = 6585.349089 days

1981
Partial Eclipse of the Moon
July 17.2036097222

Saros Number 119 (60/83)

Full Moon: July 17.19375
Earth–Moon Distance: 392,411,290.4 *m*
$\alpha = 296.°4302018$ $\delta = -20.°56020139$
$\lambda = 294.°6315999$ $\beta = +0.°6474538086$

1999
Partial Eclipse of the Moon
July 28.4908454283

Saros Number 119 (61/83)

Full Moon: July 28.4756944
Earth–Moon Distance: 393,587,859.1 *m*
$\alpha = 307.°3146657$ $\delta = -18.°28484206$
$\lambda = 305.°1437408$ $\beta = +0.°718164476$

Saros Period = 6585.2872357061 days

1981

Total Eclipse of the Sun

July 31.149697338

Saros Number 145 (20/77)

Newl Moon: July 31.1611

Earth−Moon Distance: 374,521,473.7 m

$\alpha = 130.°2480965$ $\delta = + 18.°88122653$

$\lambda = 127.°6880041$ $\beta = +0.°5520862314$

1999

Total Eclipse of the Sun

August 11.4522095833

Saros Number 145(21/77)

New Moon: August 11.463889

Earth−Moon Distance: 373,280,380.1 m

$\alpha = 140.°7659548$ $\delta = + 15.°83843963$

$\lambda = 138.°1752884$ $\beta = + 0.°479349306$

Saros Period = 6585.3025122453 days

1982
Total Eclipse of the Moon
January 9.8313469908

Saros Number 124 (47/74)

Full Moon: January 9.8284722
Earth–Moon Distance: 361,374,868.6 m
$\alpha = 110.°819049$ $\qquad$ $\delta = + 21.°76531289$
$\lambda = 109.°2739021$ $\qquad$ $\beta = - 0.°2917477576$

2000
Total Eclipse of the Moon
January 21.1993726389

Saros Number 124 (48/74)

Full Moon: January 21.19444
Earth–Moon Distance: 360,769,658.8 m
$\alpha = 122.°6275034$ $\qquad$ $\delta = + 19.°7548862$
$\lambda = 120.°4941222$ $\qquad$ $\beta = - 0.°2969776147$

Saros Period = 6585.3680256481 days

1982
Partial Eclipse of the Sun
January 25.1812809028

Saros Number 150 (15/71)

New Moon: January 25.205556
Earth–Moon Distance: 396,852,676.7 m
$\alpha = 307.°2080434$ $\delta = -20.°19318409$
$\lambda = 304.°5863967$ $\beta = -1.°109757726$

2000
Partial Eclipse of the Sun
February 5.5137345717

Saros Number 150 (16/71)

New Moon: February 5.54375
Earth–Moon Distance: 397,764,958.9 m
$\alpha = 318.°4487682$ $\delta = -17.°1866419$
$\lambda = 315.°6494401$ $\beta = -1.°091111449$

Saros Period = 6585.3324536689 days

1982
Partial Eclipse of the Sun
June 21.4941332175

Saros Number 117 (67/71)

New Moon: June 21.49444
Earth–Moon Distance: 357,327,948.2 m
$\alpha = 89.°75066305$ $\delta = +22.°19908731$
$\lambda = 89.°76909078$ $\beta = -1.°240002836$

2000
Partial Eclipse of the Sun
July 1.8125250346

Saros Number 117 (68/71)

New Moon: July 1.805556
Earth–Moon Distance: 357,375,677.2 m
$\alpha = 101.°1283697$ $\delta = +21.°73418651$
$\lambda = 100.°3309634$ $\beta = -1.°306895506$

Saros Period = 6585.3183918171 days

1982
Total Eclipse of the Moon
July 6.3131212963

Saros Number 129 (36/71)

Full Moon: July 6.3138889
Earth–Moon Distance: 405,640,146.2 m
$\alpha = 285.°1007901$ $\delta = -22.°766395$
$\lambda = 283.°8995875$ $\beta = -0.°05279949278$

2000
Total Eclipse of the Moon
July 16.5804159492

Saros Number 129 (37/71)

Full Moon: July 16.5798611
Earth–Moon Distance: 405,907,632.8 m
$\alpha = 296.°2182883$ $\delta = -212241776$
$\lambda = 294.°3195935$ $\beta = +0.°02881293502$

Saros Period = 6585.2672946529 days

1982
Partial Eclipse of the Sun
July 20.7708737267

Saros Number 155 (4/71)

New Moon: July 20.7895833
Earth–Moon Distance: 359,415,950.6 m
$\alpha = 119.°7718467$ $\delta = + 21.°93766138$
$\lambda = 117.°4329017$ $\beta = + 1.°289019783$

2000
Partial Eclipse of the Sun
July 31.077784919

Saros Number 155 (5/71)

New Moon: July 31.1006944
Earth–Moon Distance: 358,907,624.2 m
$\alpha = 130.°5766876$ $\delta = + 19.°47778064$
$\lambda = 127.°8340201$ $\beta = + 1.°208499775$

Saros Period = 6585.3069111923 days

1982
Partial Eclipse of the Sun
December 15.3825960648

Saros Number 122 (56/70)

New Moon: December 15.3875
Earth–Moon Distance: 403,741,853.3 m
$\alpha = 262.°440212$ $\delta = -22.°22739281$
$\lambda = 263.°003771$ $\beta = +1.°028486363$

2000
Partial Eclipse of the Sun
December 25.7263163658

Saros Number 122 (57/70)

New Moon: December 25.723611
Earth–Moon Distance: 403,207,360.1 m
$\alpha = 274.°6073703$ $\delta = -22.°34007834$
$\lambda = 274.°2615861$ $\beta = +1.°031079849$

Saros Period = 6585.343720301 days

1982
Total Eclipse of the Moon
December 30.4792300925

Saros Number 134 (25/73)

Full Moon: December: 30.48125
Earth–Moon Distance: 357,153,091.7 m
$\alpha = 99.°17774362$ $\delta = +23.°5568574$
$\lambda = 98.°40730246$ $\beta = +0.°3850188422$

2001
Total Eclipse of the Moon
January 9.8463626621

Saros Number 134 (26/73)

Full Moon: January 9.85
Earth–Moon Distance: 357,410,141.7 m
$\alpha = 111.°2502186$ $\delta = +22.°38105474$
$\lambda = 109.°5814624$ $\beta = +0.°3747333601$

Saros Period = 6585.3671325696 days

1983
Total Eclipse of the Sun
June 11.189871412

Saros Number 127 (56/82)

New Moon: June 11.1923611
Earth–Moon Distance: 365,707,898.7 m
$\alpha = 78.°79674841$ $\delta = + 22.°54016$
$\lambda = 79.°66197206$ $\beta = - 0.°4979737972$

2001
Total Eclipse of the Sun
June 21.4984511225

Saros Number 127 (57/82)

New Moon: June 21.498611
Earth–Moon Distance: 366,703,607.2 m
$\alpha = 90.°17564004$ $\delta = + 22.°86786114$
$\lambda = 90.°16184336$ $\beta = - 0.°5713301856$

Saros Period = 6585.3085797105 days

1983
Partial Eclipse of the Moon
June 25.3537646644

Saros Number 139 (21/82)

Full Moon: June 25.3556
Earth–Moon Distance: 400,739,892.9 m
$\alpha = 273.°5142823$ $\delta = -24.°1470666$
$\lambda = 273.°206713$ $\beta = -0.°7468869274$

2001
Partial Eclipse of the Moon
July 5.621536667

Saros Number 139 (22/82)

Full Moon: July 5.627778
Earth–Moon Distance: 399,835,984.3 m
$\alpha = 284.°8065093$ $\delta = -23.°40570025$
$\lambda = 283.°5646761$ $\beta = -0.°6609407718$

Saros Period = 6585.2677720023 days

1983
Annular Eclipse of the Sun
December 4.5135892475

Saros Number 132 (44/71)

New Moon: December 4.5180556
Earth–Moon Distance: 386,429,660.9 *m*
α = 250.°2481338   δ = − 21.°81007847
λ = 251.°7137673   β = + 0.°3840707966

2001
Annular Eclipse of the Sun
December 14.8644714004

Saros Number 132 (45/71)

New Moon: December 14.8659722
Earth–Moon Distance: 385,327,136.6 *m*
α = 262.°2924994   δ = − 22.°8575067
λ = 262.°9008251   β = − 0.°3920436225

Saros Period = 6585.3508821529 days

1983
Penumbral Eclipse of the Moon
December 20.084973206

Saros Number 144 (14/71)

Full Moon: December 20.08333
Earth–Moon Distance: 369,587,962.0 m
$\alpha = 87.°38002019 \qquad \delta = +24.°48884952$
$\lambda = 87.°6154343 \qquad \beta = +1.°07124371$

2001
Penumbral Eclipse of the Moon
December 30.4394493404

Saros Number 144 (15/71)

Full Moon: December 30.4444
Earth–Moon Distance: 370,575,442.6 m
$\alpha = 99.°55841212 \qquad \delta = +24.°20576238$
$\lambda = 98.°71265454 \qquad \beta = +1.°055100714$

Saros Period = 6585.3544761344 days

1984
Penumbral Eclipse of the Moon
Mary 15.1649155093

Saros Number 111 (65/71)

Full Moon: May 15.18680555
Earth–Moon Distance: 371,155,444.3 m
$\alpha = 232.°1319029$ $\delta = -17.°73221261$
$\lambda = 234.°2114767$ $\beta = +1.°12654294$

2002
Penumbral Eclipse of the Moon
May 26.4774681829

Saros Number 111 (66/71)

Full Moon: May 26.49375
Earth–Moon Distance: 370,048,103.3 m
$\alpha = 243.°0927189$ $\delta = -19.°92942159$
$\lambda = 244.°8152854$ $\beta = +1.°188172539$

Saros Period = 6585.3125526736 days

1984
Annular Eclipse of the Sun
May 30.7032279629

Saros Number 137 (34/70)

New Moon: May 30.7
Earth–Moon Distance: 385,341,435.6 m
$\alpha = 67.°75588026$ $\delta = + 22.°13599022$
$\lambda = 69.°47261223$ $\beta = + 0.°2679135692$

2002
Annular Eclipse of the Sun
June 10.9917931829

Saros Number 137 (35/70)

New Moon: June 10.99027778
Earth–Moon Distance: 386,661,760.4 m
$\alpha = 79.°0089242$ $\delta = + 23.°24455997$
$\lambda = 79.°91079415$ $\beta = + 0.°1893933568$

Saros Period = 6585.28856522

1984
Penumbral Eclipse of the Moon
June 13.6188554975

Saros Number 149 (1/72)
Beginning Penumbral Eclipse
(First Eclipse of the Sarose)

Full Moon: June 13.6125
Earth–Moon Distance: 381,808,687.3 m
$\alpha = 262.°1019987$ $\delta = -24.°71821278$
$\lambda = 262.°8272575$ $\beta = -1.°475360455$

2002
Penumbral Eclipse of the Moon
June 24.9016841433

Saros Number 149 (2/72)

Full Moon: June 24.90416667
Earth–Moon Distance: 380,453,006.2 m
$\alpha = 273.°4630608$ $\delta = -24.°79086293$
$\lambda = 273.°1445087$ $\beta = -1.°389365886$

Saros Period = 6585.2828286458 days

1984
Penumbral Eclipse of the Moon
November 8.7109515508

Saros Number 116 (56/73)

Full Moon: November 8.73819444
Earth–Moon Distance: 400,336,406.2 m
$\alpha = 44.°00884347$ $\delta = + 15.°69050047$
$\lambda = 46.°16825349$ $\beta = - 1.°027971681$

2002
Penumbral Eclipse of the Moon
November 20.04312620371

Saros Number 116 (57/73)

Full Moon: November 20.06527778
Earth–Moon Distance: 401,076,070.1 m
$\alpha = 55.°24059977$ $\delta = + 18.°53394419$
$\lambda = 57.°27199184$ $\beta = - 1.°04337681$

Saros Period = 6585.3321746529 days

1984
Total Eclipse of the Sun
November 22.9610417363

Saros Number 142 (21/72)

New Moon: November 22.95625
Earth–Moon Distance: 366,117,424.0 m
$\alpha = 238.°6820143$ $\delta = -20.°65302047$
$\lambda = 240.°8964111$ $\beta = -0.°3221575255$

2002
Total Eclipse of the Sun
December 4.3185288195

Saros Number 142 (22/72)

New Moon: December 4.31527778
Earth–Moon Distance: 365,242,264.1 m
$\alpha = 250.°4558289$ $\delta = -22.°53271057$
$\lambda = 252.°0010295$ $\beta = -0.°3063434071$

Saros Period = 6585.3574870832 days

1985
Total Eclipse of the Moon
May 4.8201787038

Saros Number 121 (54/84)

Full Moon: May 4.82847222
Earth–Moon Distance: 357,964,838.7 m
$\alpha = 221.°8020525 \qquad \delta = -15.°73038407$
$\lambda = 224.°1475457 \qquad \beta = +0.°37085063$

2003
Total Eclipse of the Moon
May 16.1422941435

Saros Number 121 (55/84)

Full Moon: May 16.15
Earth–Moon Distance: 357,683,196.6 m
$\alpha = 232.°5074057 \qquad \delta = -18.°53520795$
$\lambda = 234.°7529308 \qquad \beta = +0.°4344245539$

Saros Period = 6585.3221154397 days

1985
Partial Eclipse of the Sun
May 19.9238118054

Saros Number 147 (21/80)

New Moon: May 19.90347222
Earth–Moon Distance: 402,947,099.6 m
$\alpha = 56.°6203581$ $\delta = + 20.°92549671$
$\lambda = 59.°07102657$ $\beta = + 0.°9986247627$

2003
Annular Eclipse of the Sun
May 31.1931850579

Saros Number 147 (22/80)

New Moon: May 31.1805556
Earth–Moon Distance: 403,624,165.1 m
$\alpha = 67.°65063254$ $\delta = + 22.°77820189$
$\lambda = 69.°47338768$ $\beta = + 0.°9174486488$

Saros Period = 6585.2693732525 days

1985
Total Eclipse of the Moon
October 28.7220417825

Saros Number 126 (44/72)

Full Moon: October 28.734722
Earth–Moon Distance: 405,851,136.2 *m*
$\alpha = 32.°93520793$ $\delta = + 12.°86476799$
$\lambda = 35.°09139887$ $\beta = - 0.°3759131773$

2003
Total Eclipse of the Moon
November 9.03971354167

Saros Number 126 (45/72)

Full Moon: November 9.0506944
Earth–Moon Distance: 405,596,280.5 *m*
$\alpha = 43.°72554077$ $\delta = + 16.°26094098$
$\lambda = 46.°07092956$ $\beta = - 0.°4033823815$

Saros Period = 6585.3176717592 days

1985
Total Eclipse of the Sun
November 12.6176219908

Saros Number 152 (11/70)

New Moon: November 12.597222
Earth–Moon Distance: 356,879,961.2 m
$\alpha = 227.°7123503 \qquad \delta = -18.°85722925$
$\lambda = 230.°4431952 \qquad \beta = -1.°035305546$

2003
Total Eclipse of the Sun
November 23.9723475463

Saros Number 152 (12/70)

New Moon: November 23.957638889
Earth–Moon Distance: 356,811,423.1 m
$\alpha = 239.°1068305 \qquad \delta = -21.°43889884$
$\lambda = 241.°4459711 \qquad \beta = -1.°009779522$

Saros Period = 6585.3547255555 days

1986
Partial Eclipse of the Sun
April 9.2179871441

Saros Number 119 (64/71)

New Moon: April 9.255556
Earth–Moon Distance: 398,382,612.6 m
$\alpha = 17.°58400343$ $\delta = + 6.°342537879$
$\lambda = 18.°63110459$ $\beta = - 1.°035858016$

2004
Partial Eclipse of the Sun
April 19.5203701158

Saros Number 119 (65/71)

New Moon: April 19.55625
Earth–Moon Distance: 397,425,244.7 m
$\alpha = 27.°69324803$ $\delta = + 10.°23317589$
$\lambda = 29.°36429056$ $\beta = - 1.°084474303$

Saros Period = 6585.3023829717 days

1986
Total Eclipse of the Moon
April 24.5421959914

Saros Number 131 (32/72)

Full Moon: April 24.5319444
Earth–Moon Distance: 360,625,629.6 *m*
$\alpha = 211.°7936178$ $\delta = -13.°2813871$
$\lambda = 214.°1840127$ $\beta = -0.°3900998197$

2004
Total Eclipse of the Moon
May 4.8637812038

Saros Number 131 (33/72)

Full Moon: May 4.85625
Earth–Moon Distance: 361,343,090.8 *m*
$\alpha = 222.°2303751$ $\delta = -16.°58930169$
$\lambda = 224.°7944065$ $\beta = -0.°3281207459$

Saros Period = 6585.3215852124 days

1986
Annular–Total Eclipse of the Sun
October 3.7544691369

Saros Number 124 (53/73)

New Moon: October 3.78819444
Earth–Moon Distance: 374,260,436.1 m
$\alpha = 189.°3958211$ $\delta = -2.°945404991$
$\lambda = 189.°7907813$ $\beta = +1.°014685686$

2004
Partial Eclipse of the Sun
October 14.08359879629

Saros Number 124 (54/73)

New Moon: October 14.116667
Earth–Moon Distance: 375,378,498.8 m
$\alpha = 199.°4542585$ $\delta = -7.°081622806$
$\lambda = 200.°6304364$ $\beta = +1.°052044273$

Saros Period = 6585.3291296594 days

1986
Total Eclipse of the Moon
October 17.8172576011

Saros Number 136 (18/72)

Full Moon: October 17.8069444
Earth–Moon Distance: 392,877,433.9 m
$\alpha = 22.°33863047$ $\delta = + 9.°692179119$
$\lambda = 24.°24962075$ $\beta = + 0.°311313962$

2004
Total Eclipse of the Moon
October 28.1384703125

Saros Number 136 (19/72)

Full Moon: October 28.1298611
Earth–Moon Distance: 391,781,816.2 m
$\alpha = 32.°75326999$ $\delta = + 13.°49434134$
$\lambda = 35.°1348386$ $\beta = + 0.°2766631502$

Saros Period = 6585.3212127114 days

1987
Annular-Total Eclipse of the Sun
March 29.5210846875

Saros Number 129 (50/80)

New Moon: March 29.5319444
Earth-Moon Distance: 378,204,191.5 m
$\alpha = 7.°603192875$ $\delta = + 2.°946767596$
$\lambda = 8.°145307792$ $\beta = - 0.°3091073121$

2005
Annular-Total Eclipse of the Sun
April 8.8441719908

Saros Number 129 (51/80)

New Moon: April 8.85556
Earth-Moon Distance: 377,031,547.9 m
$\alpha = 17.°60367284$ $\delta = + 7.°089465966$
$\lambda = 18.°93243932$ $\beta = - 0.°3521220063$

Saros Period = 6585.3230873033 days

1987
Penumbral Eclipse of the Moon
April 14.1408472107

Saros Number 141 (22/73)

Full Moon: April 14.1048611
Earth–Moon Distance: 377,537,562.9 m
$\alpha = 201.°8912709$ $\qquad \delta = -10.°41949321$
$\lambda = 204.°1081776$ $\qquad \beta = -1.°149770502$

2005
Penumbral Eclipse of the Moon
April 24.4522157871

Saros Number 141 (23/73)

Full Moon: April 24.4208333
Earth–Moon Distance: 378,809,053.8 m
$\alpha = 212.°0845824$ $\qquad \delta = -14.°12914549$
$\lambda = 214.°736338$ $\qquad \beta = -1.°092661027$

Saros Period = 6585.3113685764 days

1987
Annular Eclipse of the Sun
September 23.1204880555

Saros Number 134 (42/71)

New Moon: September 23.1305556
Earth–Moon Distance: 395,029,808.5 m
$\alpha = 179.°5861342$ $\delta = + 0.°4747972866$
$\lambda = 179.°4314249$ $\beta = + 0.°2709947525$

2005
Annular Eclipse of the Sun
October 3.4240633679

Saros Number 134 (43/71)

New Moon: October 3.436111
Earth–Moon Distance: 396,076,417.2 m
$\alpha = 189.°4573766$ $\delta = - 3.°726931848$
$\lambda = 190.°1540275$ $\beta = + 0.°3199683871$

Saros Period = 6585.3035753124 days

1987
Penumbral Eclipse of the Moon
October 7.2090133796

Saros Number 146 (9/72)

Full Moon: October 7.175
Earth–Moon Distance: 372,002,808.8 m
$\alpha = 12.°31878886$ $\delta = + 6.°423550306$
$\lambda = 13.°83100718$ $\beta = + 1.°049272011$

2005
Partial Eclipse of the Moon
October 17.5397722571

Saros Number 146 (10/72)

Full Moon: October 17.5097222
Earth–Moon Distance: 370,876,443.7 m
$\alpha = 22.°44858269$ $\delta = + 10.°48508598$
$\lambda = 24.°64194883$ $\beta = + 1.°008824257$

Saros Period = 6585.3307588775 days

1988
Partial Eclipse of the Moon
March 3.6317159492

Saros Number 113 (62/71)

Full Moon: March 3.6673611
Earth–Moon Distance: 404,681,958.6 m
$\alpha = 164.°5693732$ $\delta = +7.°595661187$
$\lambda = 162.°8669298$ $\beta = +0.°9377177155$

2006
Penumbral Eclipse of the Moon
March 14.9445259838

Saros Number 113 (63/71)

Full Moon: March 14.982638889
Earth–Moon Distance: 405,087,775.9 m
$\alpha = 174.°6739576$ $\delta = +3.°359069064$
$\lambda = 173.°7803795$ $\beta = +0.°9682730558$

Saros Period = 6585.3128100346 days

1988
Total Eclipse of the Sun
March 18.09876865742

Saros Number 139 (28/71)

Newl Moon: March 18.084722
Earth–Moon Distance: 360,903,963.2 *m*
$\alpha = 357.°8900519$ $\delta = -0.°4315198047$
$\lambda = 357.°8924675$ $\beta = +0.°4431721139$

2006
Total Eclipse of the Sun
March 29.4397424421

Saros Number 139 (29/71)

New Moon: March 29.42708333
Earth–Moon Distance: 360,344,243.3 *m*
$\alpha = 7.°885642122$ $\delta = +3.°847108752$
$\lambda = 8.°759307655$ $\beta = +0.°4072162102$

Saros Period = 6585.3409737847 days

1988
Partial Eclipse of the Moon
August 27.4299803933

Saros Number 118 (51/75)

Full Moon: August 27.455556
Earth–Moon Distance: 357,174,306.0 m
$\alpha = 336.°2209214$ $\delta = -10.°9147745$
$\lambda = 330.°9839322$ $\beta = -0.°9303660866$

2006
Partial Eclipse of the Moon
September 7.7499687846

Saros Number 118 (52/75)

Full Moon: September 7.77916667
Earth–Moon Distance: 357,316,429.7 m
$\alpha = 346.°1533857$ $\delta = -7.°001053609$
$\lambda = 344.°5461574$ $\beta = -0.°9936777856$

Saros Period = 6585.3199883913 days

1988
Annular Eclipse of the Sun
September 11.2181114583

Saros Number 144 (15/70)

New Moon: September 11.20069444
Earth–Moon Distance: 406,360,506.3 *m*
$\alpha = 169.°5928873$ $\delta = + 4.°000686875$
$\lambda = 168.°8668781$ $\beta = - 0.°4393813457$

2006
Annular Eclipse of the Sun
September 22.5049454629

Saros Number 144 (16/70)

New Moon: September 22.489583333
Earth–Moon Distance: 406,473,992.8 *m*
$\alpha = 179.°3954908$ $\delta = - 0.°1542238197$
$\lambda = 179.°5067154$ $\beta = - 0.°3819548548$

Saros Period = 6585.2868340046 days

1989
Total Eclipse of the Moon
February 20.6374161113

Saros Number 123 (51/73)

Full Moon: February 20.647222
Earth–Moon Distance: 402,819,673.1 m

$\alpha = 153.°9615433$ $\delta = +11.°07618482$
$\lambda = 151.°8570436$ $\beta = +0.°2807431988$

2007
Total Eclipse of the Moon
March 3.9588421179

Saros Number 123 (52/73)

Full Moon: March 3.970138889
Earth–Moon Distance: 402,175,290.6 m
$\alpha = 164.°3086412$ $\delta = +7.°016754931$
$\lambda = 162.°8499072$ $\beta = +0.°30390082$

Saros Period = 6585.3214260066 days

1989
Partial Eclipse of the Sun
March 7.7979362963

Saros Number 149 (19/71)

New Moon: March 7.76319444
Earth–Moon Distance: 357,815,421.2 m
$\alpha = 348.°210764$ $\delta = -3.°791505859$
$\lambda = 347.°6788894$ $\beta = +1.°170257584$

2007
Partial Eclipse of the Sun
March 19.1479810185

Saros Number 149 (20/71)

New Moon: March 19.113194444
Earth–Moon Distance: 358,197,457.5 m
$\alpha = 358.°2953353$ $\delta = +0.°5076258528$
$\lambda = 358.°6378124$ $\beta = +1.°143757492$

Saros Period = 6585.3500447222 days

1989
Total Eclipse of the Moon
August 17.1255283912

Saros Number 128 (39/71)

Full Moon: August 17.12986111
Earth–Moon Distance: 367,595,023.7 *m*
$\alpha = 326.°4942585$ $\delta = -13.°62600619$
$\lambda = 324.°1315795$ $\beta = -0.°1571369811$

2007
Total Eclipse of the Moon
August 28.43448772

Saros Number 128 (40/71)

Full Moon: August 28.440972222
Earth–Moon Distance: 368,639,677.8 *m*
$\alpha = 336.°5945142$ $\delta = -107.°01243126$
$\lambda = 334.°6550143$ $\beta = -0.°2240456521$

Saros Period = 6585.3089593288 days

1989
Partial Eclipse of the Sun
August 31.2798955671

Saros Number 154 (5/71)

New Moon: August 31.238889
Earth–Moon Distance: 399,182,357.9 m
$\alpha = 159.°5081493$ $\delta = +7.°404347528$
$\lambda = 158.°2940478$ $\beta = -1.°137604672$

2007
Partial Eclipse of the Sun
September 11.5713011692

Saros Number 154 (6/71)

New Moon: September 11.53055556
Earth–Moon Distance: 398,225,889.7 m
$\alpha = 169.°3728084$ $\delta = +3.°398579042$
$\lambda = 168.°9002607$ $\beta = -1.°07942217$

Saros Period = 6585.2914056021 days

1990
Annular Eclipse of the Sun
January 26.7860414467

Saros Number 121 (59/71)

New Moon: January 26.805556
Earth–Moon Distance: 382,293,430.5 m
$\alpha = 308.°9438179$ $\qquad \delta = -19.°59808485$
$\lambda = 306.°3149472$ $\qquad \beta = -0.°9332907113$

2008
Annular Eclipse of the sun
February 7.1311373843

Saros Number 121 (60/71)

New Moon: February 7.155556
Earth–Moon Distance: 383,443,506.2 m
$\alpha = 320.°14425$ $\qquad \delta = -16.°52113804$
$\lambda = 317.°3975124$ $\qquad \beta = -0.°9459732957$

Saros Period = 6585.3450959376 days

1990

Total Eclipse of the Moon

February 9.8139327663

Saros Number 133 (25/71)

Full Moon: February 9.80277778

Earth–Moon Distance: 384,919,600.8 m

$\alpha = 143.°1803125$ $\delta = + 14.°13588511$

$\lambda = 140.°9234567$ $\beta = - 0.°4067516105$

2008

Total Eclipse of the Moon

February 21.1586153357

Saros Number 133 (26/71)

Full Moon: February 21.14583333

Earth–Moon Distance: 383,802,032.4 m

$\alpha = 153.°8806986$ $\delta = + 10.°38236487$

$\lambda = 152.°0307456$ $\beta = - 0.°3956615462$

Saros Period = 6585.3446825694 days

1990
Total Eclipse of the Sun
July 22.1088539005

Saros Number 126 (46/72)

New Moon: July 22.12083333
Earth–Moon Distance: 369,071,258.7 m
$\alpha = 121.°1868212$ $\quad$ $\delta = +21.°13966513$
$\lambda = 118.°8832453$ $\quad$ $\beta = +0.°7730363481$

2008
Total Eclipse of the Sun
August 1.4078834028

Saros Number 126 (47/72)

New Moon: August 1.425694444
Earth–Moon Distance: 367,952,766.1 m
$\alpha = 131.°9420553$ $\quad$ $\delta = +18.°75692771$
$\lambda = 129.°2682401$ $\quad$ $\beta = +0.°8513602369$

Saros Period = 6585.2990295023 days

1990
Partial Eclipse of the Moon
August 6.6124162846

Saros Number 138 (28/83)

Full Moon: August 6.596527778
Earth–Moon Distance: 387,839,949.6 m
$\alpha = 316.°3386679$ $\delta = -16.°01717804$
$\lambda = 314.°0582593$ $\beta = +0.°61910476$

2008
Partial Eclipse of the Moon
August 16.9027710417

Saros Number 138 (29/83)

Full Moon: August 16.886111
Earth–Moon Distance: 389,106,940.1 m
$\alpha = 326.°6648704$ $\delta = -12.°81724071$
$\lambda = 324.°5567692$ $\beta = +0.°5506093857$

Saros Period = 6585.2903547571 days

1991
Annular Eclipse of the Sun
January 15.9885437154

Saros Number 131 (50/71)

New Moon: January 15.99305556
Earth–Moon Distance: 401,407,383.9 m
$\alpha = 297.°2810082 \qquad \delta = -21.°33402093$
$\lambda = 295.°2741962 \qquad \beta = -0.°2565472087$

2009
Annular Eclipse of the Sun
January 26.3238769213

Saros Number 131 (51/71)

New Moon: January26.3298611111
Earth–Moon Distance: 402,088,015.1 m
$\alpha = 308.°8684887 \qquad \delta = -18.°92633367$
$\lambda = 306.°4139253 \qquad \beta = -0.°2650758988$

Saros Period = 6585.3353332059 days

1991
Penumbral Eclipse of the Moon
January 30.2794918171

Saros Number 143 (17/73)

Full Moon: January 30.25694444
Earth–Moon Distance: 364,893,991.3 m
$\alpha = 132.°3009521$ $\delta = + 16.°62770911$
$\lambda = 130.°1658793$ $\beta = - 1.°109453708$

2009
Penumbral Eclipse of the Moon
February 9.6448693288

Saros Number 143 (18/73)

Full Moon: February 9.61736111
Earth–Moon Distance: 364,126,762.6 m
$\alpha = 143.°3991238$ $\delta = + 13.°32443003$
$\lambda = 141.°3837428$ $\beta = - 1.°108047523$

Saros Period = 6585.3653775117 days

1991
Penumbral Eclipse of the Moon
June 27.1194528704

Saros Number 110 (70/72)

Full Moon: June 27.1236111
Earth-Moon Distance: 406,231,720.6 m
$\alpha = 275.°4264594$ $\delta = -24.°62490351$
$\lambda = 274.°9328975$ $\beta = -1.°278485835$

2009
Penumbral Eclipse of the Moon
July 7.3754021991

Saros Number 110 (71/72)

Full Moon: July 7.389583333
Earth-Moon Distance: 406,138,790.0 m
$\alpha = 286.°6886237$ $\delta = -23.°92603462$
$\lambda = 285.°2225024$ $\beta = -1.°364343157$

Saros Period = 6585.2559493287 days

A complete description of Saros series 136 is presented by Mark Littmann, Fred Espenak, and Ken Willcox.[16] Solar Saros 136 series has 71 members. The entire series lasts 1262 years. The longest eclipse occurred on June 20, 1955, with a maximum duration of totality at 7 minutes, 8 seconds.

Saros_136_animation.gif

Saros 136 animation from Wikipedia, the free encyclopedia.

1991
Total Eclipse of the Sun
July 11.7959272683

Saros Number 136 (36/71)

New Moon: July 11.79583333
Earth–Moon Distance: 357,736,093.7 *m*
$\alpha = 119.°5424229 \qquad \delta = + 22.°09487349$
$\lambda = 108.°9735398 \qquad \beta = - 0.°001503697309$

2009
Total Eclipse of the Sun
July 22.1062584259

Saros Number 136 (37/71)

New Moon: July 22.107638889
Earth–Moon Distance: 357,529,648.9 *m*
$\alpha = 121.°5868983 \qquad \delta = + 20.°34599717$
$\lambda = 119.°4137141 \qquad \beta = + 0.°07410076305$

Saros Period = 6585.3103311576 days

1991
Penumbral Eclipse of the Moon
July 26.7958988775

Saros Number 148 (2/71)

Full Moon: July 26.766667
Earth–Moon Distance: 404,048,576.6 m
$\alpha = 305.°5920425$ $\delta = -18.°04909639$
$\lambda = 303.°6088481$ $\beta = +1.°334116731$

2009
Penumbral Eclipse of the Moon
August 6.07287126158

Saros Number 148 (3/71)

Full Moon: August 6.03819444
Earth–Moon Distance: 404,577,129.5 m
$\alpha = 316.°2096193$ $\delta = -15.°37717537$
$\lambda = 314.°1232767$ $\beta = +1.°267804746$

Saros Period = 6585.27697238408 days

1991
Partial Eclipse of the Moon
December 21.4332659258

Saros Number 115 (56/72)

Full Moon: December 21.432638889
Earth-Moon Distance: 359,225,738.5 *m*
$\alpha = 88.°94870279$ $\delta = + 24.°42978984$
$\lambda = 89.°04269186$ $\beta = + 0.°9939933036$

2009
Partial Eclipse of the Moon
December 31.7949774883

Saros Number 115 (57/72)

Full Moon: December 31.800694444
Earth-Moon Distance: 359,734,022.3 *m*
$\alpha = 101.°1287177$ $\delta = + 24.°05283217$
$\lambda = 100.°153243$ $\beta = + 1.°004930127$

Saros Period = 6585.3617115625 days

1992
Annular Eclipse of the Sun
January 4.9685590163

Saros Number 141 (22/70)

New Moon: January 4.96527778
Earth–Moon Distance: 405,716,629.7 m
$\alpha = 285.°0407582$ $\delta = -22.°34586238$
$\lambda = 283.°8879397$ $\beta = +0.°3712117775$

2010
Annular Eclipse of the Sun
January 15.3057896991

Saros Number 141 (23/70)

New Moon: January 15.299305556
Earth–Moon Distance: 405,402,361.3 m
$\alpha = 296.°963469$ $\delta = -20.°75360384$
$\lambda = 295.°0880037$ $\beta = +0.°3679891398$

Saros Period = 6585.3372306827 days

1992
Partial Eclipse of the Moon
June 15.2033734491

Saros Number 120 (57/84)

Full Moon: June 15.20138889
Earth–Moon Distance: 429,057,575.3 m
$\alpha = 263.°8250346$ $\delta = -23.°90037834$
$\lambda = 264.°3560386$ $\beta = -0.°581986444$

2010
Partial Eclipse of the Moon
June 26.4773375463

Saros Number 120 (58/84)

Full Moon: June 26.47916667
Earth–Moon Distance: 395,101,993.4 m
$\alpha = 275.°1838731$ $\delta = -24.°01543414$
$\lambda = 274.°7343788$ $\beta = -0.°6612977482$

Saros Period = 6585.2739640972 days

1992
Total Eclipse of the Sun
June 30.5162382988

Saros Number 146 (26/76)

New Moon: June 30.5125
Earth–Moon Distance: 362,516,062.9 m
$\alpha = 99.°7283944$ $\delta = + 22.°37386164$
$\lambda = 98.°99051709$ $\beta = - 0.°7623242621$

2010
Total Eclipse of the Sun
July 11.8270219675

Saros Number 146 (27/76)

New Moon: July 11.8194444
Earth–Moon Distance: 363,322,441.5 m
$\alpha = 110.°9904621$ $\delta = + 21.°33610909$
$\lambda = 109.°4926468$ $\beta = - 0.°6939767787$

Saros Period = 6585.3107836687 days

1992
Total Eclipse of the Moon
December 9.9886998033

Saros Number 125 (47/72)

Full Moon: December 9.986805556
Earth–Moon Distance: 375,453,713.7 m
$\alpha = 77.°13304478$ $\delta = + 23.°21942662$
$\lambda = 78.°19078527$ $\beta = + 0.°3060700501$

2010
Total Eclipse of the Moon
December 21.3427925695

Saros Number 125 (48/72)

Full Moon: December 21.34236111
Earth–Moon Distance: 376,530,236.7 m
$\alpha = 89.°27581744$ $\delta = + 23.°75025084$
$\lambda = 89.°33714166$ $\beta = + 0.°3126271354$

Saros Period = 6585.3540927662 days

1992
Partial Eclipse of the Sun
December 24.03135621528

Saros Number 151 (13/72)

New Moon: December 24.02986111
Earth–Moon Distance: 391,523,933.9 m
$\alpha = 272.°6758632$ $\delta = -22.°41360578$
$\lambda = 272.°4739647$ $\beta = +1.°002707389$

2011
Partial Eclipse of the Sun
January 4.3855620139

Saros Number 151 (14/72)

New Moon: January 4.37708333
Earth–Moon Distance: 390,508,685.0 m
$\alpha = 284.°8203067$ $\delta = -21.°72715258$
$\lambda = 283.°7480738$ $\beta = +1.°007607267$

Saros Period = 6585.3542058611 days

1993
Partial Eclipse of the Sun
May 21.6067805671

Saros Number 118 (67/72)

New Moon: May 21.5875
Earth–Moon Distance: 391,230,983.2 m
$\alpha = 58.°37477312$ $\delta = +21.°32953822$
$\lambda = 60.°7563085$ $\beta = +1.°043498453$

2011
Partial Eclipse of the Sun
June 1.89025708333

Saros Number 118 (68/72)

New Moon: June 1.87708333
Earth–Moon Distance: 392,467,808.3 m
$\alpha = 69.°46565021$ $\delta = +23.°22413464$
$\lambda = 71.°19145045$ $\beta = +1.°11566589$

Saros Period = 6585.283476516 days

1993
Total Eclipse of the Moon
June 4.5417264467

Saros Number 130 (33/72)

Full Moon: June 4.54305556
Earth–Moon Distance: 375,749,737.2 m
$\alpha = 252.°5409002$ $\delta = -22.°3104$
$\lambda = 253.°8849909$ $\beta = +0.°1574563179$

2011
Total Eclipse of the Moon
June 15.8424603471

Saros Number 130 (34/72)

Full Moon: June 15.84305556
Earth–Moon Distance: 374,506,278.1 m
$\alpha = 263.°8783264$ $\delta = -23.°23127226$
$\lambda = 264.°3763353$ $\beta = +0.°08859317291$

Saros Period = 6585.3007339004 days

1993
Partial Eclipse of the Sun
November 13.9188871642

Saros Number 123 (52/70)

New Moon: November 13.8986111
Earth–Moon Distance: 361,889,172.8 m
$\alpha = 229.°1140126$ $\delta = -19.°21192347$
$\lambda = 231.°815064$ $\beta = -1.°027519621$

2011
Partial Eclipse of the Sun
November 25.2717649537

Saros Number 123 (53/70)

New Moon: November 25.2569444
Earth–Moon Distance: 361,189,857.2 m
$\alpha = 240.°5527692$ $\delta = -21.°75138108$
$\lambda = 242.°8259622$ $\beta = -1.°047144084$

Saros Period = 6585.3528777898 days

1993
Total Eclipse of the Moon
November 29.2659651736

Saros Number 135 (22/71)

Full Moon: November 29.271527778
Earth–Moon Distance: 396,372,380.7 m
$\alpha = 65.°21544015$ $\delta = +21.°11672209$
$\lambda = 66.°97922468$ $\beta = -0.°3636396043$

2011
Total Eclipse of the Moon
December 10.6057901967

Saros Number 135 (23/71)

Full Moon: December 10.608333
Earth–Moon Distance: 397,269,515.6 m
$\alpha = 77.°1356403$ $\delta = +22.°55371372$
$\lambda = 78.°13420329$ $\beta = -0.°3572382483$

Saros Period = 6585.339825023 days

1994
Annular Eclipse of the Sun
May 10.7221025346

Saros Number 128 (57/73)

New Moon: May 10.71319444
Earth–Moon Distance: 405,511,130.2 *m*
$\alpha = 47.°36011439$ $\delta = + 18.°06127031$
$\lambda = 49.°90754691$ $\beta = + 0.°3583222557$

2012
Annular Eclipse of the Sun
May 20.9994103588

Saros Number 128 (58/73)

New Moon: May 20.9909722221
Earth–Moon Distance: 405,849,376.8 *m*
$\alpha = 58.°17401502$ $\delta = + 20.°65690256$
$\lambda = 60.°43228742$ $\beta = + 0.°4249536845$

Saros Period = 6585.2773078242 days

1994
Partial Eclipse of the Moon
May 25.1400828357

Saros Number 140 (24/80)

Full Moon: May 25.15208333
Earth–Moon Distance: 359,762,163.5 m
$\alpha = 241.°6848257$ $\delta = -19.°97962791$
$\lambda = 243.°5236532$ $\beta = +0.°894819967$

2012
Partial Eclipse of the Moon
June 4.4593861225

Saros Number 140 (25/80)

Full Moon: June 4.46667
Earth–Moon Distance: 359,244,896.1 m
$\alpha = 252.°8750396$ $\delta = -2`.°66484457$
$\lambda = 254.°116329$ $\beta = +0.°8351288943$

Saros Period = 6585.3193032868 days

1994
Total Eclipse of the Sun
November 3.5743851967

Saros Number 133 (44/72)

New Moon: November 3.56597222
Earth-Moon Distance: 357,408,132.3 m
$\alpha = 218.°4772527$ $\delta = -15.°464795$
$\lambda = 221.°0160632$ $\beta = -0.°3498197418$

2012
Total Eclipse of the Sun
November 13.9292242013

Saros Number 133 (45/72)

New Moon: November 13.92222
Earth-Moon Distance: 357,611,305.0 m
$\alpha = 229.°5203256$ $\delta = -18.°63405238$
$\lambda = 232.°0364325$ $\beta = -0.°3697468662$

Saros Period = 6585.3548390046 days

1994
Penumbral Eclipse of the Moon
November 18.267496088

Saros Number 145 (10/71)

Full Moon: November 18.28958333
Earth–Moon Distance: 406,345,725.7 m
$\alpha = 53.°34223153$ $\delta = + 18.°17403887$
$\lambda = 55.°43559831$ $\beta = - 0.°9750762143$

2012
Penumbral Eclipse of the Moon
November 28.5999105904

Saros Number 145 (11/71)

Full Moon: November 28.61527778
Earth–Moon Distance: 406,346,692.3 m
$\alpha = 64.°91472395$ $\delta = + 20.°45582012$
$\lambda = 66.°59119747$ $\beta = - 0.°9680302252$

Saros Period = 6585.3324145024 days

1995
Partial Eclipse of the Moon
April 15.532460625

Saros Number 112 (64/72)

Full Moon: April 15.5055556
Earth–Moon Distance: 364,307,419.9 m
$\alpha = 203.°237023$ $\delta = -10.°70736502$
$\lambda = 205.°4445874$ $\beta = -0.°9314883267$

2013
Partial Eclipse of the Moon
April 25.8567787733

Saros Number 112 (65/72)

Full Moon: April 25.83125
Earth–Moon Distance: 365,242,688.6 m
$\alpha = 213.°4696399$ $\delta = -14.°48884375$
$\lambda = 216.°1216598$ $\beta = -0.°9827230728$

Saros Period = 6585.3243181483 days

112 The Dance of the Moon

The Dance of the Moon

1995
Annular Eclipse of the Sun
April 29.7244674304

Saros Number 138 (30/70)

New Moon: April 29.73333
Earth–Moon Distance: 401,841,940.5 m
$\alpha = 36.°53458614$ $\delta = +14.°15726346$
$\lambda = 38.°82134963$ $\beta = -0.°2982932441$

2013
Annular Eclipse of the Sun
May 10.01365625

Saros Number 138 (31/70)

New Moon: May 10.0194444
Earth–Moon Distance: 401,046,999.3 m
$\alpha = 47.°05885183$ $\delta = +17.°35823156$
$\lambda = 49.°44126249$ $\beta = -0.°2410139761$

Saros Period = 6585.28918882 days

1995
Penumbral Eclipse of the Moon
October 8.6965308333

Saros Number 117 (51/72)

Full Moon: October 8.66111
Earth–Moon Distance: 387,036,927.6 m
$\alpha = 13.°73653522$ $\delta = + 6.°983315279$
$\lambda = 15.°34619672$ $\beta = + 1.°019531714$

2013
Penumbral Eclipse of the Moon
October 19.01854950231

Saros Number 117 (52/72)

Full Moon: October 18.98472222
Earth–Moon Distance: 385,832,352.8 m
$\alpha = 23.°89647044$ $\delta = + 11.°09311706$
$\lambda = 26.°18832519$ $\beta = + 1.°054547915$

Saros Period = 6585.3220186687 days

1995
Total Eclipse of the Sun
October 24.1822960648

Saros Number 143 (23/73)

New Moon: October 24.1916667
Earth–Moon Distance: 369,550,779.7 m
$\alpha = 208.°172521$ $\delta = - 11.°20934746$
$\lambda = 210.°1483032$ $\beta = + 0.°3360856886$

2013
Total Eclipse of the Sun
November 3.5269269558

Saros Number 143 (24/73)

New Moon: November 3.5347222
Earth–Moon Distance: 370,528,132.4 m
$\alpha = 218.°8162477$ $\delta = - 14.°87260159$
$\lambda = 221.°1430805$ $\beta = + 0.°3147163251$

Saros Period = 6585.344630891 days

Variation of Period

𝒯ℎℯarithmetic mean of the 67 Saros periods is 6585.32016681 ± 0.0035048024 standard error. A graph of the variation in the length of Saros is shown in figure 3.3.

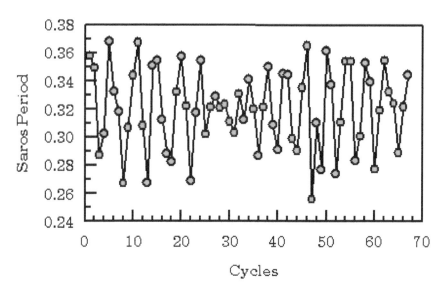

Figure 3.3
Variation in the length of the 67 Saros cycles.

The variation of the 33 Saros periods of the Solar eclipses is shown in Figure 3.4. The variation of the 34 Saros periods of the Lunar eclipses is shown in Figure 3.5.

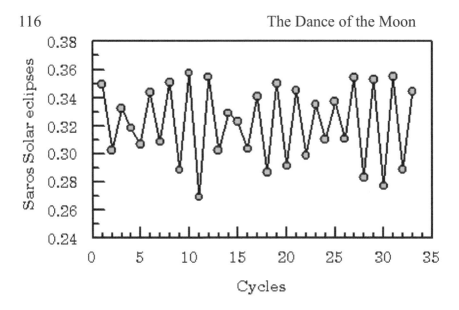

Figure 3.4. Variation of the 33 Saros periods of the Solar eclipses.

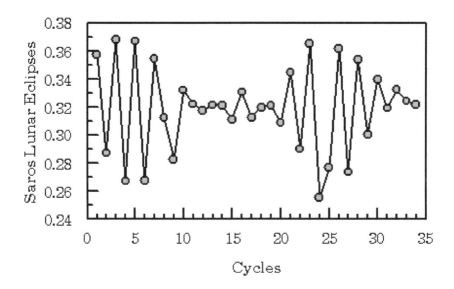

Figure 3.5. Variation of the 34 Saros periods of the Lunar eclipses.

References

1. O. Neugebauer, *The Exact Sciences in Antiquity* (New York: Dover Publications, Inc. 1969), pp. 141–142.

2. William F. Rigge, "The Lunar Saros," *Popular Astronomy,* **26** (1918), pp. 85–95 at p.95.

3. George O. Abell, David Morrison, Sidney C. Wolff, *Exploration of the Universe* (Saunders College Publishing, 1991), p. 100.

4. Mark Littmann, Fred Espenak, Ken Willcox, *Totality Eclipses of the Sun* (Oxford University Press, Updated Third Edition, 2009), p.26.

5. Ibid., p. 20.

6. A. E. Roy, "The Use of the Saros in Lunar Dynamical Studies," *The Moon,* **7** (1973), pp. 6–13.

7. Ettore Perozzi, Archie E. Roy, Bonnie A. Steves, Giovannie B. Valsecchi, "Significant high number commensurabilities in the main lunar problem. I-The Saros as a near–periodicity of the moon's orbit," *Celestial Mechanics and Dynamical Astronomy,* **52** (1991), pp. 241–261.

8. Jean Meeus, *More Mathematical Astronomy Morsels* (Richmond, VA: Willmann–Bell, Inc. 2002), p. 56.

9. Martin C. Gutzwiller, "Moon–Earth–Sun: The oldest three–body problem," *Reviews of Modern Physics,* **70** (1998), pp. 589–639 at p.596.

10. Jean Meeus, *Mathematical Astronomy Morsels III* (Richmond, VA: Willmann–Bell, Inc. 2004), pp. 87–113.

11. Bruce McCurdy, "Orbital Oddities," *Journal of the Royal Astronomical society of Canada,* **105** (2011), pp. 120–122.

12. Saros in Encyclopedia Britannica.

13. William F. Rigge, *op. cit.,* p. 93.

14. Fred Espenak, *Fifty Year Canon of Solar Eclipses: 1986–2035* (Cambridge, Massachusetts: Sky Publishing Corporation 1987).

15. Fred Espenak, *Fifty Year Canon of Lunar Eclipses: 1986–2035* (Cambridge, Massachusetts: Sky Publishing Corporation 1989).

16. Mark Littmann, Fred Espenak, Ken Willcox, *op. cit.,* pp. 225–235.

CHAPTER 4

Perturbation by the Sun

*Isaac*Newton in *principia* considered perturbation of the Moon's motion by the Sun in two places:[1] Corollaries 2 through 11 of proposition 66 of book 1,[2] and proposition 25 of book 3.[3]

The position of the Sun at various Earth–Moon distances for the years 1981, 1982, 1983, 1984, 1985, 1986, 1987, and 1997 are presented in Figures 4.1, 4.2, 4.3, 4.4, 4.5, 4.6, 4.7, and 4.8.

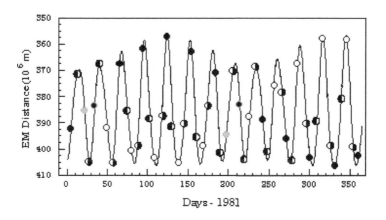

Figure 4.1
The position of the Sun at various Earth–Moon distances in 1981.

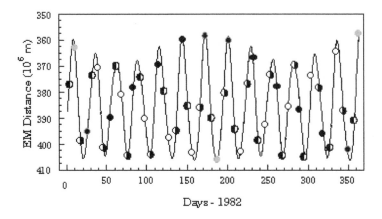

Figure 4.2
The position of the Sun at various Earth–Moon distances in 1982.

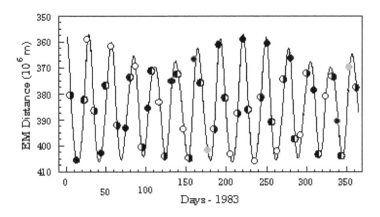

Figure 4.3
The position of the Sun at various Earth–Moon distances in 1983.

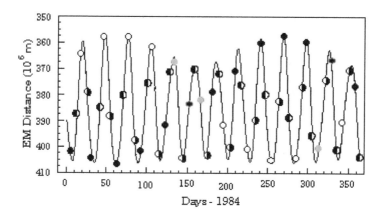

Figure 4.4
The position of the Sun at various Earth–Moon distances in 1984.

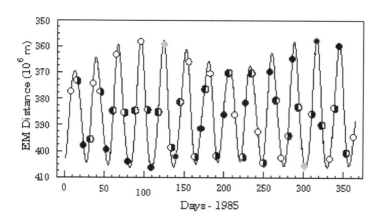

Figure 4.5
The position of the Sun at various Earth–Moon distances in 1985.

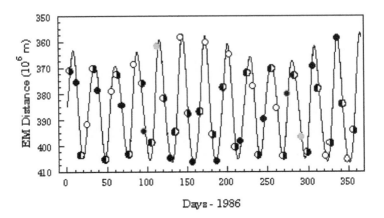

Figure 4.6
The position of the Sun at various Earth–Moon distances in 1986.

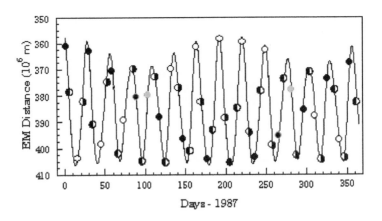

Figure 4.7
The position of the Sun at various Earth–Moon distances in 1987.

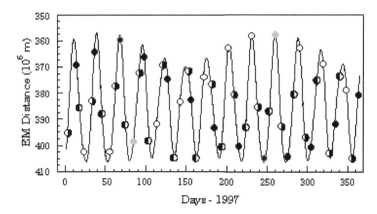

Figure 4.8
The position of the Sun at various Earth–Moon distances in 1997

.

        The position of the Sun at various velocities of the motion of the
Moon for the years 1981, 1982, 1983, 1984, 1985, 1986, 1987, and 1997
are presented in Figures 4.9, 4.10, 4.11, 4.12, 4.13, 4.14, 4.15, and 4.16.

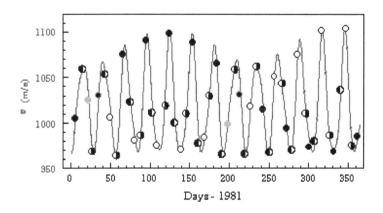

Figure 4.9
The position of the Sun at various velocities of the motion of the Moon in
1981.

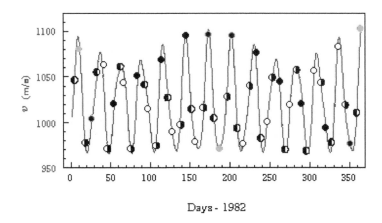

Figure 4.10
The position of the Sun at various velocities of the motion of the Moon in 1982.

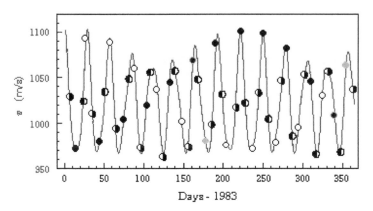

Figure 4.11
The position of the Sun at various velocities of the motion of the Moon in 1983.

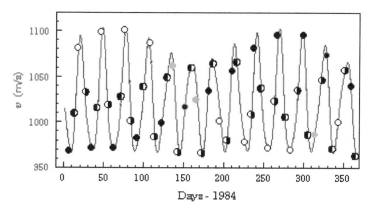

Figure 4.12
The position of the Sun at various velocities of the motion of the Moon in
1984.

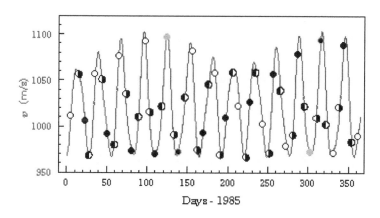

Figure 4.13
The position of the Sun at various velocities of the motion of the Moon in
1985.

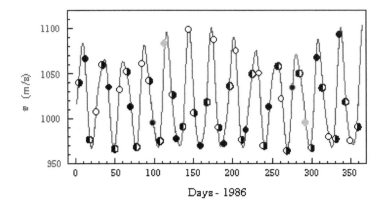

Figure 4.14
The position of the Sun at various velocities of the motion of the Moon in 1986.

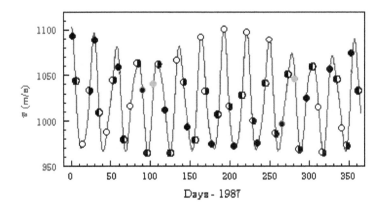

Figure 4.15
The position of the Sun at various velocities of the motion of the Moon in 1987.

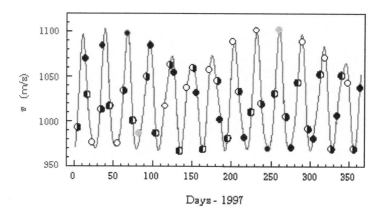

Figure 4.16
The position of the Sun at various velocities of the motion of the Moon in
1997.

The Sun's position at various Earth's gravitational force attraction
of the Moon in 1981 and 1997 are presented in Figures 4.17 and 4.18.

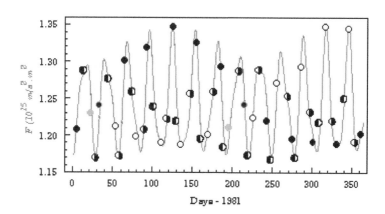

Figure 4.17
The Sun's position at various Earth's gravitational force attraction of the
Moon in 1981.

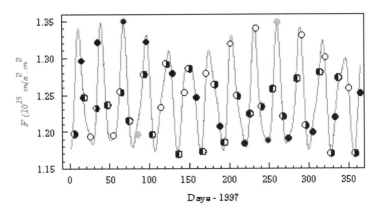

Figure 4.18
The Sun's position at various Earth's gravitational force attraction of the Moon in 1997.

The Sun's position at various products of velocity and distance, *vr,* for the years 1981, 1982, 1983, 1984, 1985, 1986, 1987, and 1997 are presented in Figures 4.19, 4.20, 4.21, 4.22, 4.23, 4.24, 4.25, and 4.26.

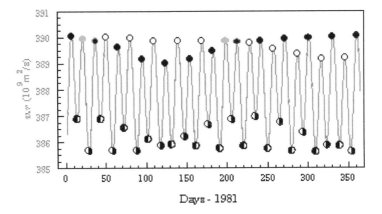

Figure 4.19
The Sun's position at various magnitudes of *vr* in 1981.

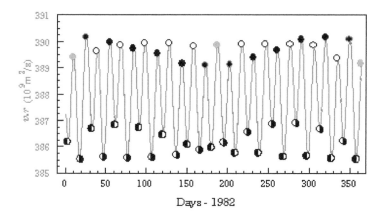

Figure 4.20
The Sun's position at various magnitudes of vr in 1982.

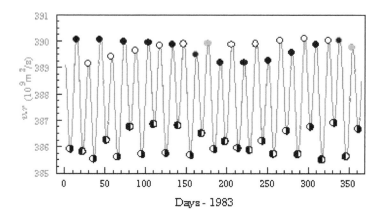

Figure 4.21
The Sun's position at various magnitude of vr in 1983.

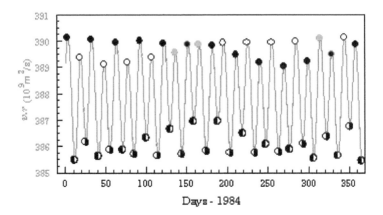

Figure 4.22
The Sun's position at various magnitudes of vr in 1984.

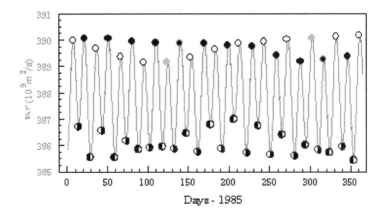

Figure 4.23
The Sun's position at various magnitudes of vr in 1985.

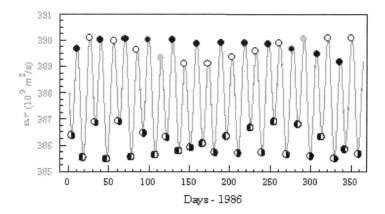

Figure 4.24

The Sun's position at various magnitudes of vr in 1986.

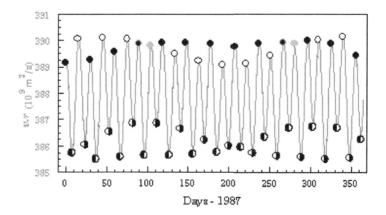

Figure 4.25

The Sun's position at various magnitudes of vr in 1987.

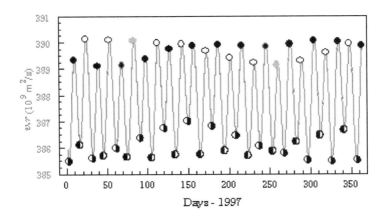

Figure 4.26
The Sun's position at various magnitudes of *vr* in 1997.

The position of the Sun at various latitudes in the years 1981, 1982, 1987, and 1997 are presented in Figures 4.27, 4.28, 4.29, and 4.30.

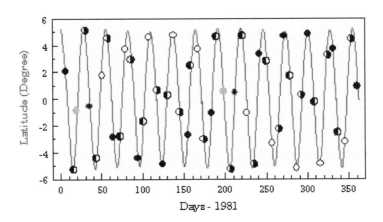

Figure 4.27
The Sun's position at various latitudes in 1981.

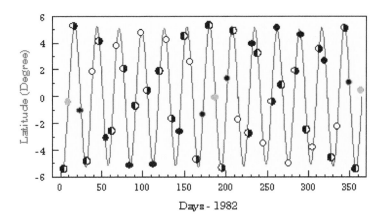

Figure 4.28
The Sun's position at various latitudes in 1982.

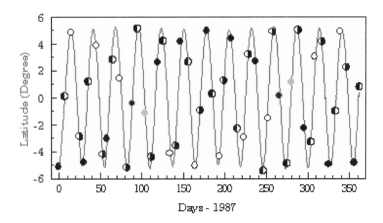

Figure 4.29
The Sun's position at various latitudes in 1987.

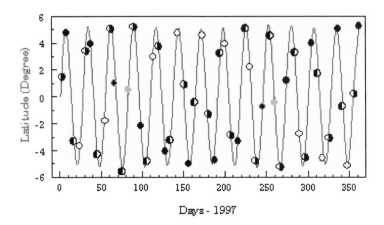

Figure 4.30
The Sun's position at various latitudes in 1997.

The position of the Sun at various Declinations in the years 1981, 1982, 1987, and 1997 are presented in Figures 4.31, 4.32, 4.33, and 4.34.

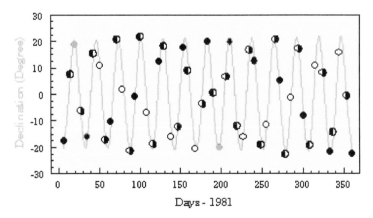

Figure 4.31
The Sun's position at various declinations in 1981.

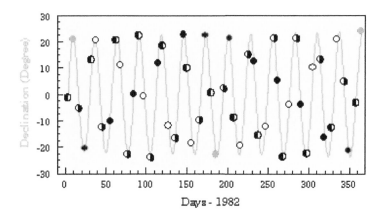

Figure 4.32
The Sun's position at various declinations in 1982.

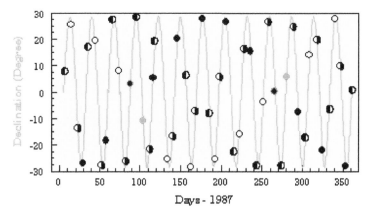

Figure 4.33
The Sun's position at various declinations in 1987.

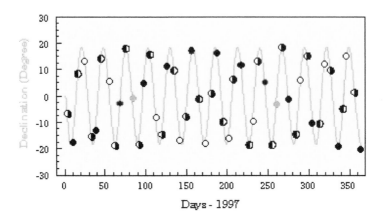

Figure 4.34
The Sun's position at various declinations in 1997.

Angular Momentum

$\mathcal{A}ngular$ momentum of an orbiting planet or satellite is:[4]

$$L = mr^2\omega = mr^2\frac{v}{r} = mvr \qquad\qquad 4.1$$

Angular momentum is a vector. The cross or vector product of the two vectors r and v is another vector perpendicular to the plane formed by the two vectors, see Figure 4.35.

$$v \times r = n[v][r]\sin\theta \qquad\qquad 4.2$$

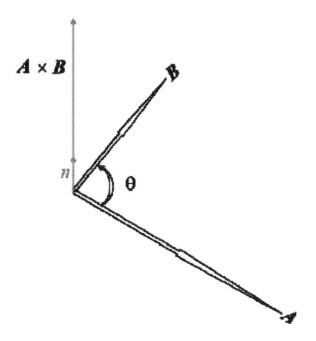

Figure 4.35

The vector product (or cross product) $A \times B = AB \sin\theta$ is another vector perpendicular to the plane formed by A and B and is equal to the product of the magnitude of the two vectors multiplied by the sine of the smaller angle between them. n is a vector of unit length perpendicular to the plane of A and B and so directed that a right-handed screw rotated from A toward B will advance in the direction of n.

     The graphs of vr at equatorial Ascending Nodes (AN) and Descending Nodes (DN) versus Declination Maximum or Minimum and versus Latitude at equatorial Maximum or Minimum are presented in the next 8 Figures. The calculations of the velocity at the Ascending Node (AN) and at the Descending Node (DN) of the Declinations were made by the distance traveled between plus and minus half–a–day from the nodes. The angle θ between r and v is 90°, and the $\sin\theta = 1$.

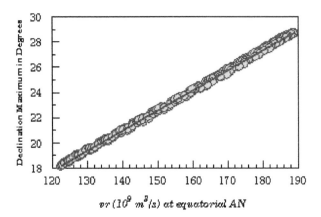

Figure 4.36

The magnitude of vr at equatorial AN plotted versus Declination Maximum. The plot represents 401 data from 1981 to 2010. The equation of the least squares line of regression is: Declination $= 0.159\ vr - 1.3$. The correlation coefficient is 0.9996.

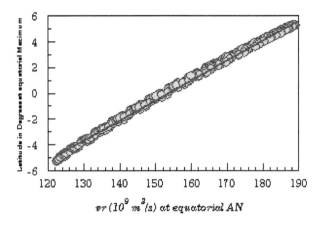

Figure 4.37

The magnitude of vr at equatorial AN plotted versus Latitude at equatorial Maximum. The plot represents 401 data from 1981 to 2010. The equation of the least squares line of regression is: Latitude $= 0.156\ vr - 24$. The correlation coefficient is 0.9986.

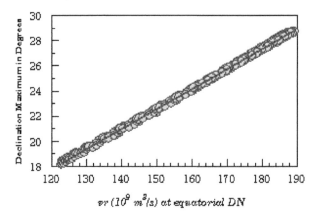

Figure4.38

The magnitude of vr at equatorial DN plotted versus Declination Maximum. The plot represents 401 data from 1981 to 2010. The equation of the least squares line of regression is: Declination = $0.159\,vr - 1.3$. The correlation coefficient is 0.9996.

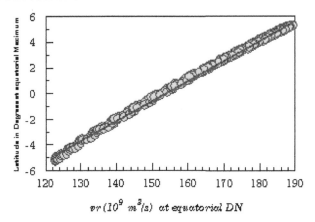

Figure 4.39

The magnitude of vr at equatorial DN plotted versus Latitude at equatorial Maximum. The plot represents 401 data from 1981 to 2010. The equation of the least squares line of regression is: Latitude = $0.156\,vr - 24$. The correlation coefficient is 0.9986.

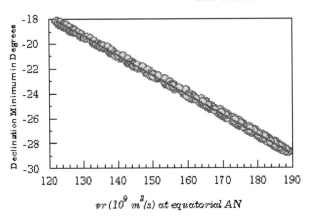

Figure 4.40
The magnitude of vr at equatorial AN plotted versus Declination Minimum. The plot represents 401 data from 1981 to 2010. The equation of the least squares line of regression is: Declination $= -0.159\, vr + 1.32$. The correlation coefficient is 0.9996.

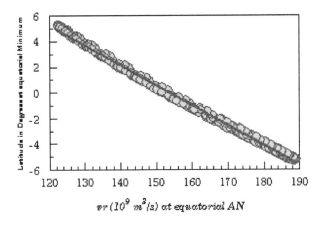

Figure 4.41
The magnitude of vr at equatorial AN plotted versus Latitude at equatorial Minimum. The plot represents 400 data from 1981 to 2010. The equation of the least squares line of regression is: Latitude $= -0.156\, vr + 24$. The correlation coefficient is 0.9982.

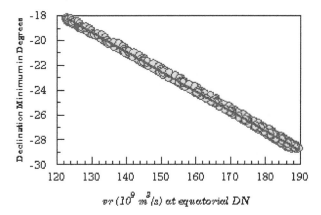

Figure 4.42

The magnitude of vr at equatorial DN plotted versus Declination Minimum. The plot represents 401 data from 1981 to 2010. The equation of the least squares line of regression is: Declination = $-0.159\ vr + 1.32$. The correlation coefficient is 0.9991

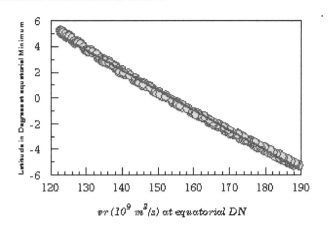

Figure 4.43

The magnitude of vr at equatorial DN plotted versus Latitude at equatorial Minimum. The plot represents 400 data from 1981 to 2010. The equation of the least squares line of regression is: Latitude = $-0.156\ vr + 24$. The correlation coefficient is 0.9986.

The vertical angles up to ± 5.°30 of the latitudes of the Moon's orbit are caused by the Sun.

References

1. I. Bernard Cohen and Anne Whitman, *Isaac Newton THE PRINCIPIA Mathematical Principles of Natural Philosophy,* preceded by *A Guide to Newton's Principia* by I. Bernard Cohen (Berkeley, Los Angeles, London: University of California Press,1999), p. 257.

2. Ibid., pp. 570–579.

3. Ibid., pp. 839–840.

4. Pari Spolter, "Kepler's Second Law and Conservation of Angular Momentum," *Physics Essays,* **24** (2011), pp. 260–266. Also see http://www.youtube.com/watch?v=tYoe8KdU21A
http://parispolter.com/

CHAPTER 5

Regression of the Nodes

$\mathcal{I}_n$ chapter 2, pages 16–17, the length of the months for the Ascending Node (AN) to AN and Descending Node (DN) to DN of the Latitudes are given. Figure 5.1 presents the length of the months for the AN to AN and for the DN to DN of the Declination in the interval from 1981 to 2008, 373 cycles. The arithmetic mean for AN to AN is 27.3190 days ± 0.00387 standard errors. The arithmetic mean for DN to DN is 27.3167 ± 0.00365 standard errors.

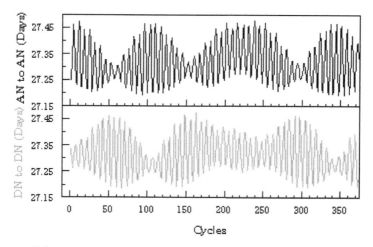

Figure 5.1
The period of AN to AN and DN to DN of the Declination in days.

The periods from the AN to DN and from the DN to AN of the Latitudes are presented in Figure 5.2, and these periods for the Declinations are presented in Figure 5.3.

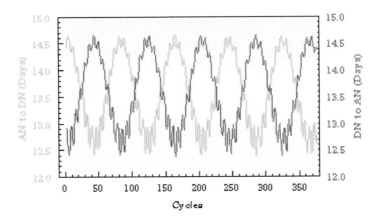

Figure 5.2
The periods in days of the AN to DN and the DN to AN of the Latitudes.

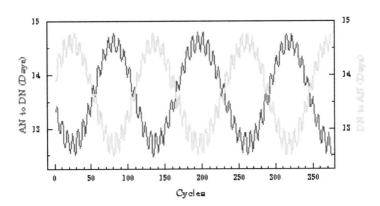

Figure 5.3
The periods in days of the AN to DN and the DN to AN of the declinations.

The Maximums and Minimums of the declination and of the latitude for this period are shown in Figures 5.4 and 5.5.

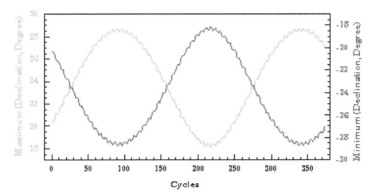

Figure 5.4
The Maximum and Minimum of declination for 373 cycles.

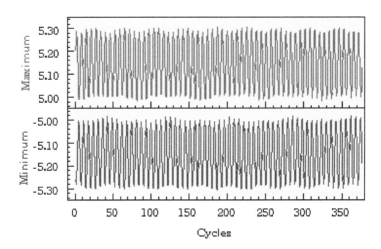

Figure 5.5
The Maximum and Minimum of latitude in degree for 373 cycles.

The orbit of the Moon in the equatorial and ecliptic coordinate systems is shown in Figure 5.6.

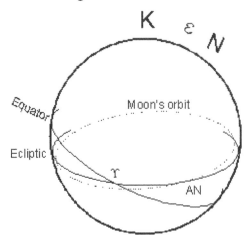

Figure 5.6 The orbit of the Moon.

The latitudes and declinations of the orbit of the Moon for the years 1981, 1982, 1987, 1992, 1997, 2001, and 2006 are presented in Figures 5.7, 5.8, 5.9, 5.10, 5.11, 5.12, and 5.13.

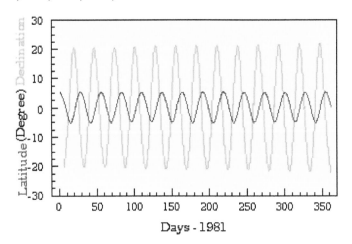

Figure 5.7
The latitudes and declinations of the orbit of the Moon for the year 1981.

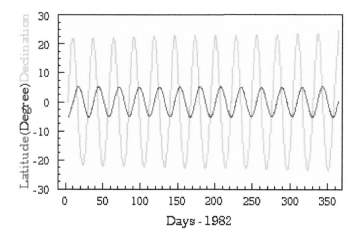

Figure 5.8
The latitudes and declinations of the orbit of the Moon for the year 1982.

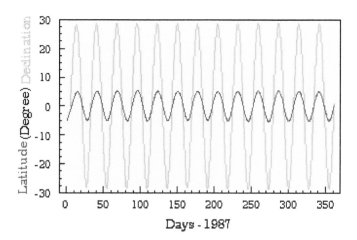

Figure 5.9
The latitudes and declinations of the orbit of the Moon for the year 1987.

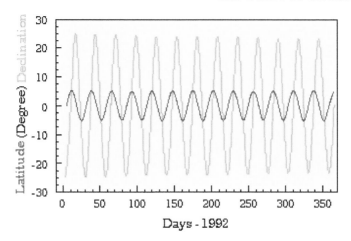

Figure 5.10
The latitudes and declinations of the orbit of the Moon for the year 1992.

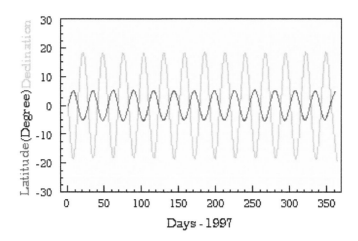

Figure 5.11
The latitudes and declinations of the orbit of the Moon for the year 1997

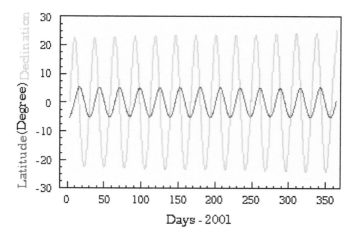

Figure 5.12
The latitudes and declinations of the orbit of the Moon for the year 2001.

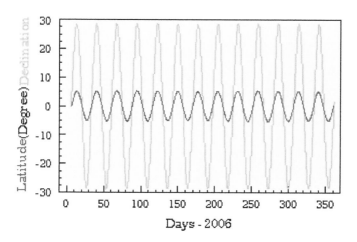

Figure 5.13
The latitudes and declinations of the orbit of the Moon for the year 2006

The Moon's Right Ascensions in equatorial coordinate system at the Descending Node (DN) and at the Ascending Node (AN) from 1981 to 2010, 401 cycles, are presented in Figure 5.14, and the Longitudes in ecliptic coordinate system at DN and at AN from 1981 to 2008, 376 cycles, are presented in Figure 5.15.

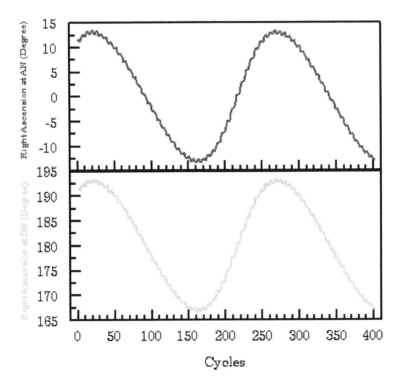

Figure 5.14
The Right Ascension of the Moon's orbit in equatorial coordinate system at the DN and AN for 401 cycles.

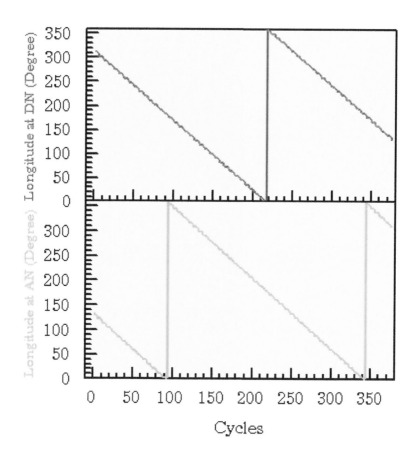

Figure 5.15

The Moon's Longitudes in ecliptic coordinate system at DN and at AN from 1981 to 2008, 376 cycles.

The next two figures present regression of the nodes and of the maximums and minimums of the declinations from 1981 to 2008, 373 cycles.

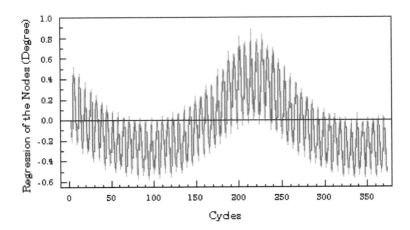

Figure 5.16
Regression of the DN and of the Maximums of the Declination for 373
cycles.

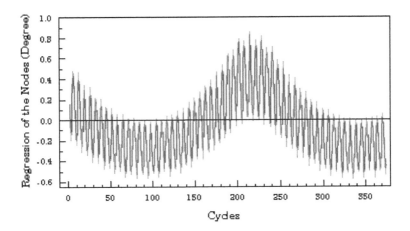

Figure 5.17
Regression of the AN and of the Minimums of the Declination for 373
cycles.

The next figure presents regression of the nodes of the orbit of the Moon's latitudes from 1981 to 2008, 376 cycles.

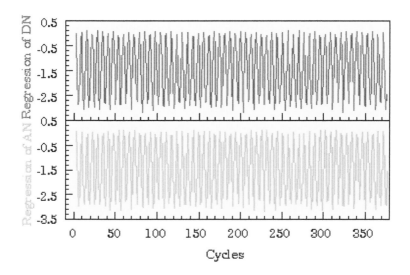

Figure 5.18
Regression of the AN and DN in degrees of the latitude of the Moon in 376 cycles.

George Elwood Smith, the 2009 Nobel laureate in physics, points out:[1]

The lines of nodes ...vary from their mean motions in complex ways, so that the location of a node can deviate in either direction from the mean motion by as much as 1°.40′.

Archie Edminston Roy (1924–2012), Professor of Astronomy at the University of Glasgow, and author of the book *Orbital Motion* states:[2]

The line of nodes regresses in the plane of the ecliptic, making one revolution in 6798.3 days (about 18.6 years).

Melek and Esat Hamzaoglu[3] present data in graphical form based on their calculations of the equations, and not on the Astronomical Almanac's lunar polynomial tables. They give the rotation period of the regression of the line of nodes as 6797.72 days (18.61 years).

Regression of the AN and of the DN making one revolution in the plane of the ecliptic in days for 126 cycles is presented in figure 5.19.

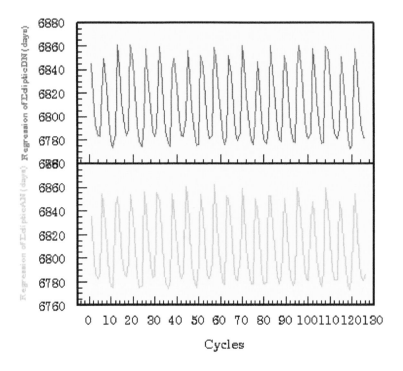

Figure 5.19

Regression of the AN and of the DN making one revolution in the plane of the ecliptic in days for 126 cycles.

For the AN the period of one revolution varied from 6772.63 to 6862.29 days with the arithmetic mean at 6808.84 days ± 2.59 standard error. For the DN the period of one revolution varied between 6772.34 to 6860.92 days with the arithmetic mean at 6809.41 days ± 2.62 standard error.

References

1. I. Bernard Cohen and Anne Whitman, *Isaac Newton THE PRINCIPIA Mathematical Principles of Natural Philosophy,* preceded by *A Guide to Newton's Principia* by I. Bernard Cohen (Berkeley, Los Angeles, London: University of California Press,1999), pages 252–253.

2.Archie Edminston Roy, *Orbital Motion* (Bristol and Philadelphia: Institute of Physics Publishing, 1988), Third edition, pages 280–281.

3. Melek Hamzaoglu and Ezat Hamzaoglu, "Graphical Representation of Periodic Motion of the Line of Nodes of the Moon." *Earth, Moon, and Planets,* **60** (1993), pp. 233–250.

CHAPTER 6

Advance of the Perigee

The variation of the Perigee to Perigee and of the Apogee to Apogee periods in days from 1981 to 2008, 371 cycles, are shown in Figure 2.4 on page 16, chapter 2 of this book. Perigee to Perigee varies from 24.65095 to 28.55735 days, a change of 3.9064 days. Apogee to Apogee varies from 26.98475 to 27.8958 days, a change of only 0.91105 days.

The variation of the distances of the Perigees and Apogees for this interval is shown in figure 6.1.

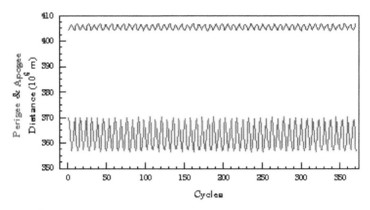

Figure 6.1
Perigee to Perigee and Apogee to Apogee distances for 28 years.

A more detailed of this variation for a shorter period of time is shown in figure 6.2, and compared with Earth Sun distances of the Perihelion and Aphelion shown in figure 6.3.

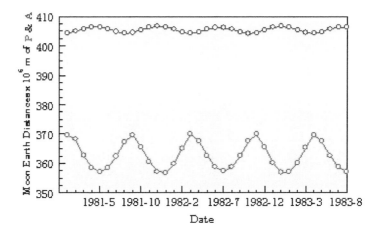

Figure 6.2
Perigee to Perigee and Apogee to Apogee distances for 3 years.

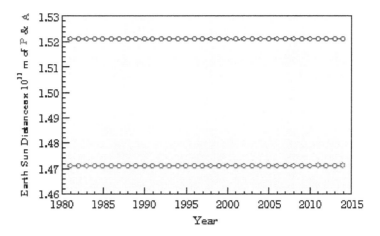

Figure 6.3
Earth Sun distances of the Perihelion and the Aphelion for 35 years.

The Earth Sun distances and the Earth Moon distances in one year are shown in the next figure.

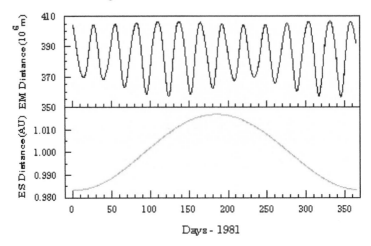

Figure 6.4 The Earth Sun and the Earth Moon distances in 1981.

Gravitational force of the Sun affects the position of the Moon as shown in the next 8 figures.

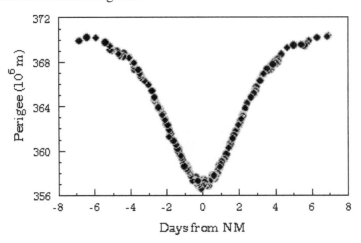

Figure 6.5 Distances of the Perigee at days from the New Moon.

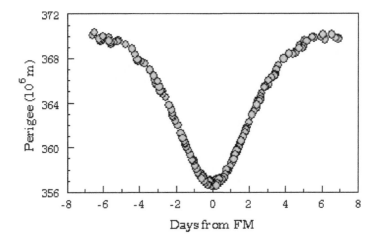

Figure 6.6
Distances of the Perigee at days from the Full Moon.

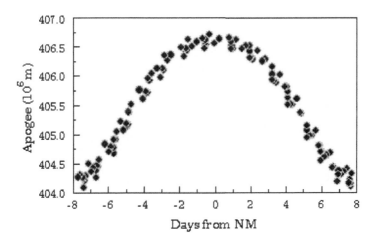

Figure 6.7
Distances of the Apogee at days from the New Moon.

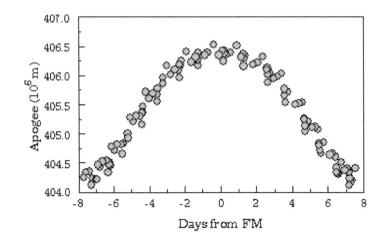

Figure 6.8
Distances of the Apogee at days from the Full Moon.

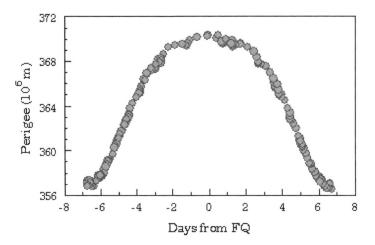

Figure 6.9
Distances of the Perigee at days from the First Quarter.

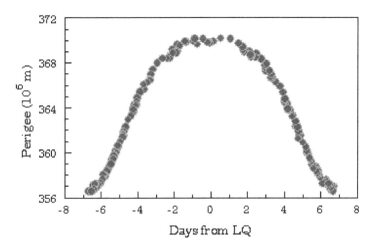

Figure 6.10
Distances of the Perigee at days from the Last Quarter.

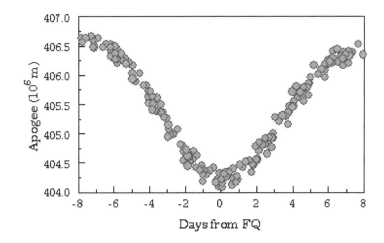

Figure 6.11
Distances of the Apogee at days from the First Quarter.

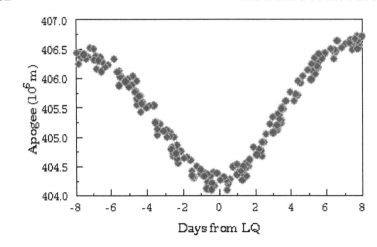

Figure 6.12
Distances of the Apogee at days from the Last Quarter.

As the Sun pulls the Moon close to the Earth, the Moon's gravitational attraction of the Earth increases. The next figure shows the advance and regression of the Perigee and Apogee.

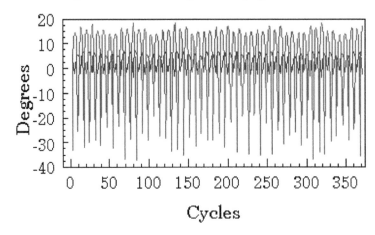

Figure 6.13
Advance and regression of Perigee to Perigee and of Apogee to Apogee in degrees from 1981 to 2008, 371 cycles.

At the closest distances to the Earth, perigee to perigee advances. At the most-distant distances it regresses. Perigees advanced as much as +18°.19 and regressed as much as −36°.97 in one revolution during this interval as shown in the next figure.

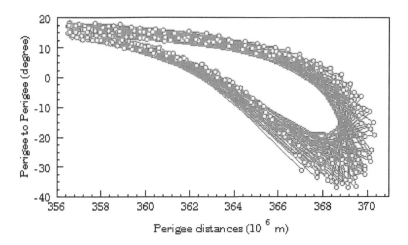

Figure 6.14
Advance and regression of the Perigee at various distances from the Earth.

On page 220 of *Gravitational Force of the Sun*[1] and on page 43 of "New Concepts in Gravitation"[2] three possible explanations for the unaccounted perihelion advance of the planet Mercury were discussed. We now add the greater velocity of Mercury compared to the other planets.

The Apogee, being farther away from the Earth advances and regresses much less than the perigee, as shown in figure 6.13 above, but as shown in the next figure it advances at closer distances to the Earth and regresses at the farther distances.

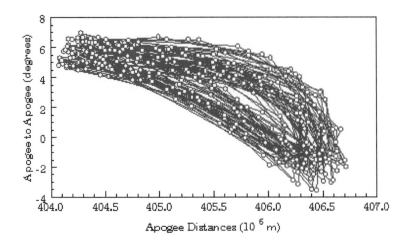

Figure 6.15
Advance and regression of the Apogee at various distances from the Earth.

When Perigee advances the Apogee regresses, and when Perigee regresses the Apogee advances, as shown in the next figure.

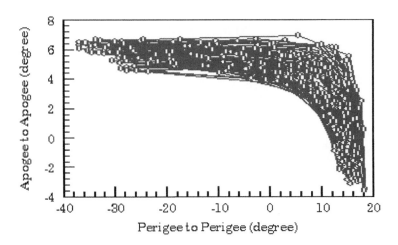

Figure 6.16
Advance and regression of Perigee and Apogee in one revolution.

The advance in degrees of the Perigees and of the Apogees in one year from 1981 to 2008 is shown in the next figure.

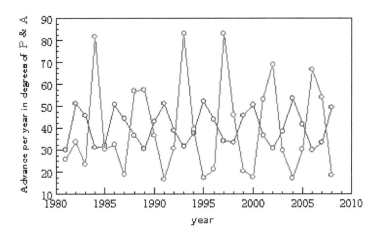

Figure 6.17
The advance of the Perigee and of the Apogee per year in degrees for 28 years.

For the Apogee the minimum value is 30°.188, the maximum value is 53°.546, and the arithmetic mean is 40°.452 ± 1°.515 standard errors. For the Perigee the minimum value is 16°.929, the maximum value is 83°.255, and the arithmetic mean is 39°.821 ± 4°.092 standard errors.

A comparison of the wide variations of the advance of the Moon's perigee with the modest changes of the Earth's perihelion is presented in the next figure.

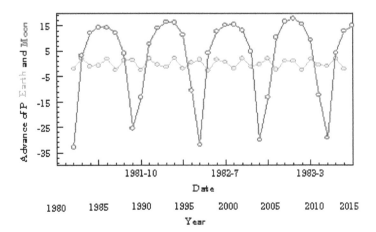

Figure 6.18 Advance and regression of the Moon's Perigee for three years, and of the Earth's Perihelion for 35 years in degrees.

When the Perigee is at the closest distance to the Earth, the Apogee is at the farthest distance, and when the Perigee is at the farthest distance, the Apogee is at the closest distance to the Earth, as shown in the next figure.

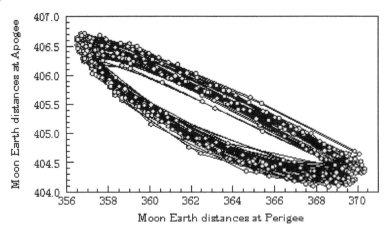

Figure 6.19
Moon Earth distances at Perigee and Apogee times $10^6 m$.

Another peculiarity of the Moon's orbit is that there is no straight line of apsides in most revolutions. The Apogee and Perigee move in different ways. The next two figures show the Perigee to Apogee in degrees at different distances of the Perigees and Apogees.

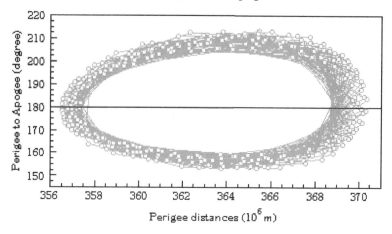

Figure 6.20

Perigee to Apogee in degrees at various distances of the Perigee.

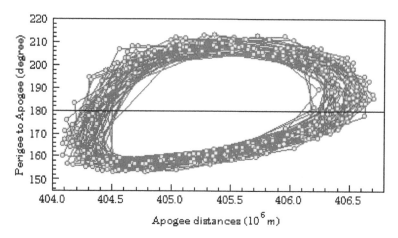

Figure 6.21

Perigee to Apogee in degrees at various distances of the Apogee.

The Perigee to Apogee in degrees of the orbit of the Moon around the Earth for three years is compared with the Perihelion to Aphelion in degrees of the orbit of the Earth around the Sun for 35 years in the next figure.

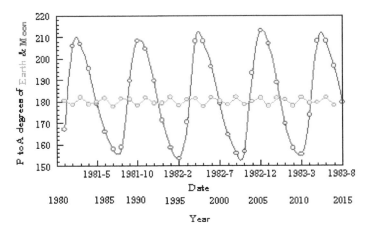

Figure 6.22
Perigee to Apogee in degrees of the orbit of the Moon around the Earth for three years, and Perihelion to Aphelion in degrees of the orbit of the Earth around the Sun for 35 years.

On page 27 of *Newton's Forgotten Lunar Theory,* Nicholas Kollerstrom points out[3]

The line [of apside] is a mathematical abstraction, because the true apogee and perigee positions can deviate widely from it. Perigee swings nearly 30° to and fro, twice a year. The apogee and perigee positions oscillate in rather different ways, and do not remain opposite each other in the sky.

Of the 371 revolutions of Perigees from 1981 to 2008, only 24 at the closest distances of the Perigees to the Earth, and only 5 at the farthest distances had the Perigee to Apogee within 180° ± 5°. The eccentricity calculations for these two groups are shown in the next two figures.

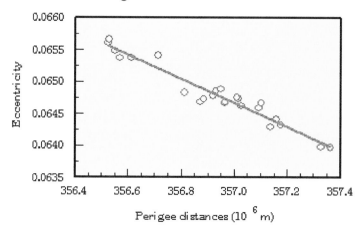

Figure 6.23
Orbital eccentricity of the Moon at 24 closest perigee distances.

The orbital eccentricity is higher at the closest distances of the Perigee. The equation of the least squares line of regression is:
$e = -0.00188 \, P + 0.735$. The correlation coefficient is 0.97.

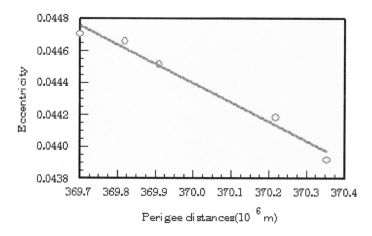

Figure 6.24
Orbital eccentricity of the Moon at 5 farthest perigee distances.

The orbital eccentricity is higher at the closest distances of the Perigee. The equation of the least squares line of regression is:
$e = -0.00121 \, P + 0.492$. The correlation coefficient is 0.99

Data presented in this chapter clearly demonstrate that what causes the complex orbital motions is the gravitational attraction of the Earth by the Moon as it comes close to it.

References

1. Pari Spolter, *Gravitational Force of the Sun* (Granada Hills, California: Orb Publishing Company, 1930, p. 220.

2. Pari Spolter, "New Concepts in Gravitation," *Physics Essays,* **18** (2005), pp. 37–49.

3. Nicholas Kollerstrom, *Newton's Forgotten Lunar Theory* (Santa Fe, New Mexico: Green Lion Press, 2000), p. 27.

CHAPTER 7

The Obliquity of the Ecliptic

Earth doesn't orbit the Sun upright. It is tilted on its axis by 23°.43928.[1] The Earth's orbital plane around the Sun is known as the ecliptic plane, and the Earth's tilt is known as the obliquity of the ecliptic. The angle between the ecliptic and the celestial equator is denoted by ε. The axis of the Earth remains oriented in the same direction with reference to the background stars at various positions in its annual path around the Sun as shown in the next figure.

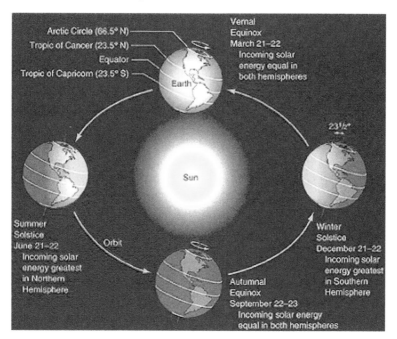

Figure 7.1
The obliquity of the ecliptic. Image courtesy of NASA.

The intervals in degrees between the right ascensions of the various phases of the Moon in 1981 are shown in the next figure.

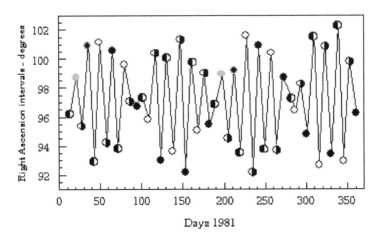

Figure 7.2
The Right Ascension intervals between various phases of the Moon in the year 1981.

Vernal Equinox is at day 79.71, Summer Solstice at 172.49, Autumnal Equinox at 266.13, and Winter Solstice at 355.95. Between Winter Solstice and Vernal Equinox, and also between Summer Solstice and Autumnal Equinox, the intervals in degrees between the first or last quarters and the new moons or full moons were greater and the intervals between the new moons or full moons and the first or last quarters smaller. The reverse was observed between Vernal Equinox and Summer Solstice, and also between Autumnal Equinox and Winter Solstice.

This serendipitous discovery was so surprising and important that I found it necessary to confirm the phenomenon for every year from 1981 to 2010. The graphs for these thirty years are presented in this chapter.

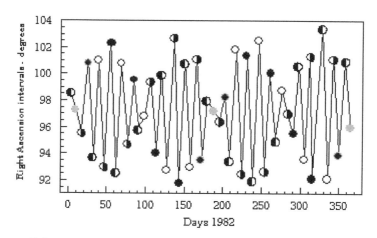

Figure 7.3
The Right Ascension intervals between various phases of the Moon in the
year 1982.

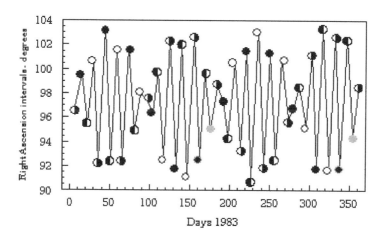

Figure 7.4
The Right Ascension intervals between various phases of the Moon in the
year 1983.

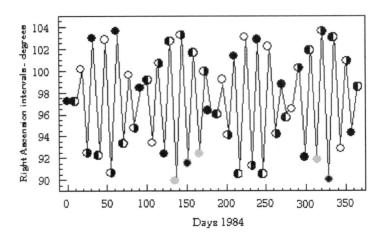

Figure 7.5
The Right Ascension intervals between various phases of the Moon in the
year 1984.

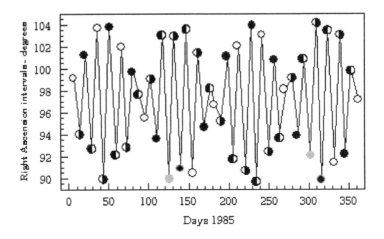

Figure 7.6
The Right Ascension intervals between various phases of the Moon in the
year 1985.

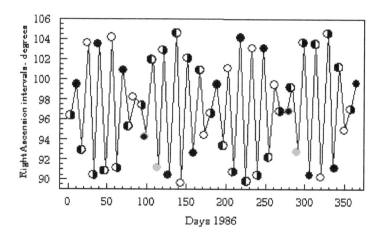

Figure 7.7
The Right Ascension intervals between various phases of the Moon in the year 1986.

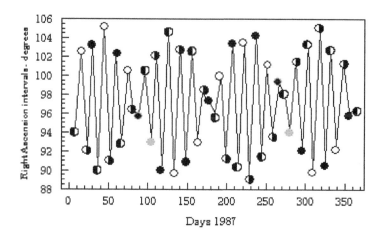

Figure 7.8
The Right Ascension intervals between various phases of the Moon in the year 1987

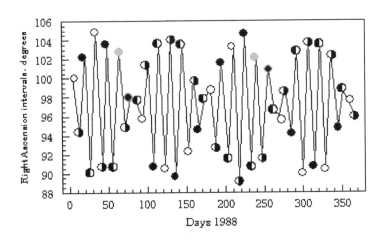

Figure 7.9
The Right Ascension intervals between various phases of the Moon in the
year 1988

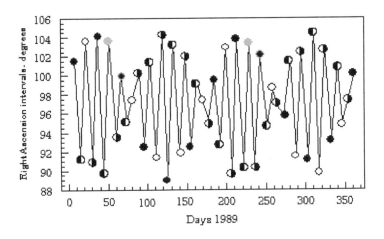

Figure 7.10
The Right Ascension intervals between various phases of the Moon in the
year 1989.

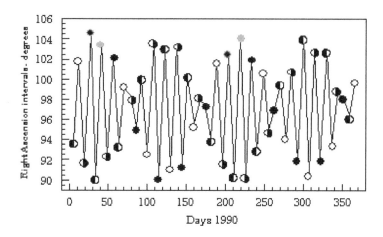

Figure 7.11
The Right Ascension intervals between various phases of the Moon in the year 1990.

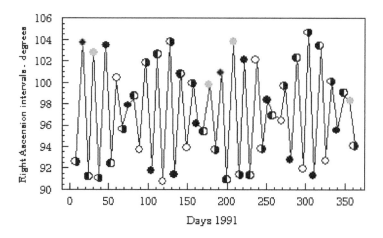

Figure 7.12
The Right Ascension intervals between various phases of the Moon in the year 1991.

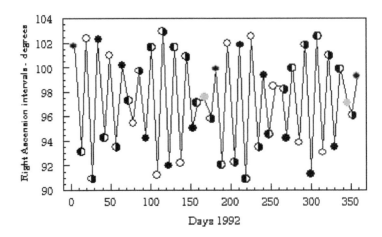

Figure 7.13
The Right Ascension intervals between various phases of the Moon in the year 1992.

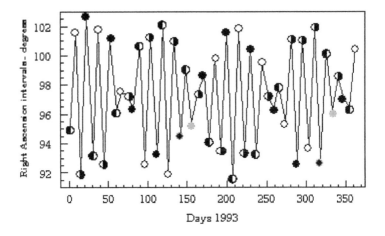

Figure 7.14
The Right Ascension intervals between various phases of the Moon in the year 1993.

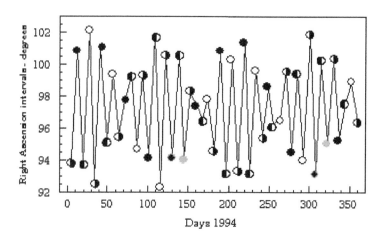

Figure 7.15
The Right Ascension intervals between various phases of the Moon in the year 1994.

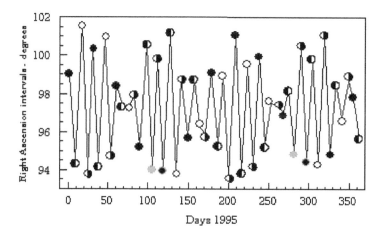

Figure 7.16
The Right Ascension intervals between various phases of the Moon in the year 1995.

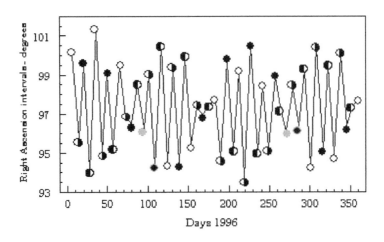

Figure 7.17
The Right Ascension intervals between various phases of the Moon in the year 1996.

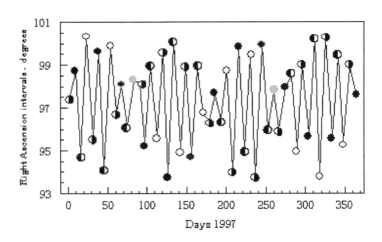

Figure 7.18
The Right Ascension intervals between various phases of the Moon in the year 1997.

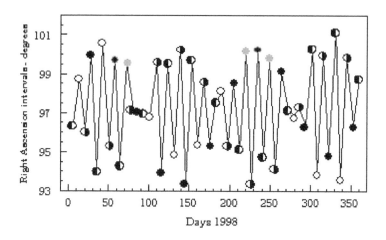

Figure 7.19
The Right Ascension intervals between various phases of the Moon in the year 1998.

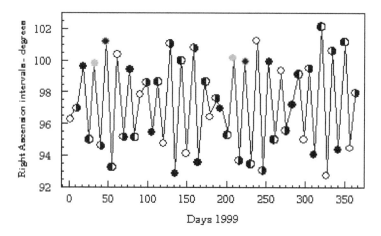

Figure 7.20
The Right Ascension intervals between various phases of the Moon in the year 1999.

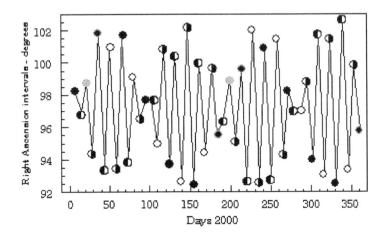

Figure 7.21
The Right Ascension intervals between various phases of the Moon in the
year 2000.

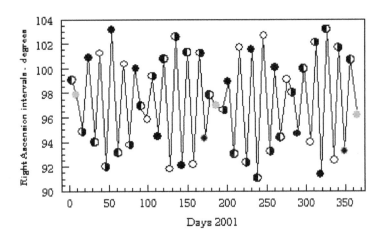

Figure 7.22
The Right Ascension intervals between various phases of the Moon in the
year 2001.

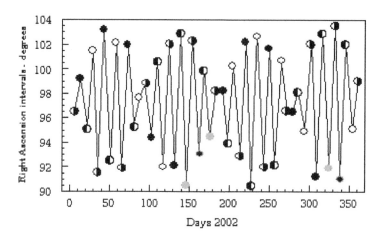

Figure 7.23
The Right Ascension intervals between various phases of the Moon in the year 2002.

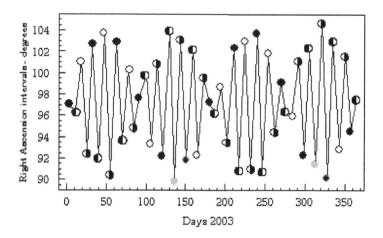

Figure 7.24
The Right Ascension intervals between various phases of the Moon in the year 2003.

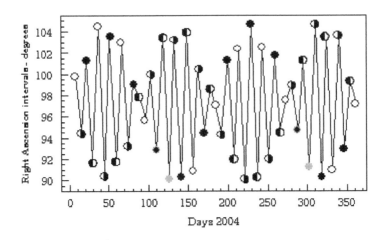

Figure 7.25
The Right Ascension intervals between various phases of the Moon in the year 2004.

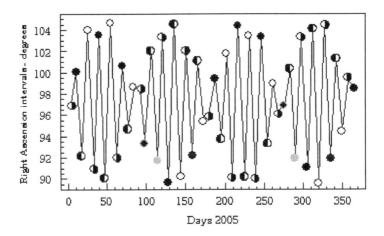

Figure 7.26
The Right Ascension intervals between various phases of the Moon in the year 2005.

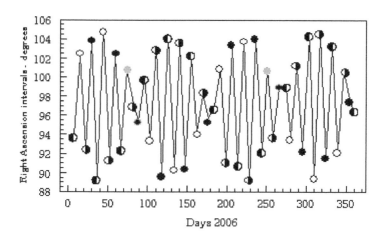

Figure 7.27
The Right Ascension intervals between various phases of the Moon in the
year 2006.

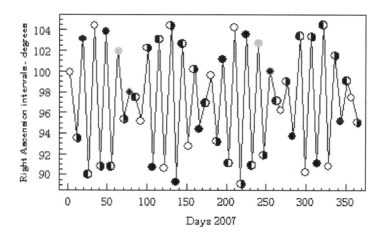

Figure 7.28
The Right Ascension intervals between various phases of the Moon in the
year 2007.

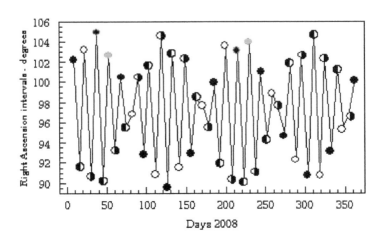

Figure 7.29
The Right Ascension intervals between various phases of the Moon in the year 2008.

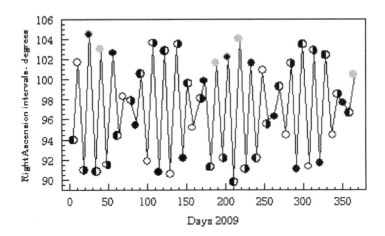

Figure 7.30
The Right Ascension intervals between various phases of the Moon in the year 2009.

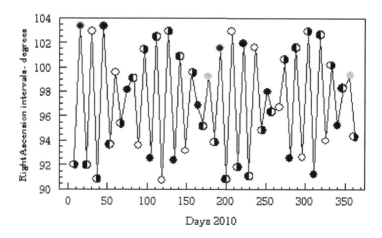

Figure 7.31
The Right Ascension intervals between various phases of the Moon in the year 2010.

The obliquity of the ecliptic slowly decreases with time.[2] The equation for the change in the obliquity of the ecliptic with respect to the mean equator of date is given in the *Explanatory Supplement to the Astronomical Almanac* 2013[3]

$$\varepsilon_A = 23° \ 26' \ 21".406 - 46".836 \ 769 \ T - 0".0001831 \ T^2 + 0".00200340 \ T^3 - 0".576 \times 10^{-6} \ T^4 - 4".34 \times 10^{-8} \ T^5$$

where T is time interval in Julian centuries from J2000.0.

For the 30 years reported in this chapter the variation is not significant.

References

1. *The Astronomical Almanac* for the year 2015 (Washington, D.C. US Government Printing Office), p. K7.

2. James G. Williams, "Contributions to the Earth's Obliquity Rate, Precession, and Nutation," *the Astronomical Journal,* **108** (1994), pp. 711–724.

3. C. Y. Hohenkerk, in *Explanatory Supplement to the Astronomical Almanac,* 3rd Edition, Edited by Sean E. Urban and P. Kenneth Seidelmann (University Science Books, Mill Valley, California, 2013), p. 274.

Index

Advance of the perigee156–170
Advance of the perihelion of
    the planet Mercury 163
Air Force Cambridge Research
    Laboratories Lunar Ranging
    Observatory 3
Aldrin, Edwin 3
Angular momentum 3,
    136–142
Anomalistic month 13, 15, 18
Apache Point Observatory
    Lunar Laser–ranging
    Operation (APOLLO
    acronym) 6
Apollo 11 3, 4, 5
Apollo 14 4, 5
Apollo 15 4, 5
Apside, line of 167–168
Armstrong, Neil 3
Astronomical Almanac 6
Atomic clock vii, 3, 6
Babylonian astronomers 44
Beethoven, Ludwig van
    The Moonlight Sonata viii
Calculations 3, 6–9
Canons of solar and lunar
    eclipses 46

Centre d'étude et de recherche
    en géodynamique et
    astronomie (CERGA) 6
Cesium atomic clock 6
Chaldeans 44
Coincidence of periods 17–43
Corner reflectors 3, 4, 5
Daily velocity of the Moon 7
Deferents 1
Diacu,Florin 2–3
Draconic month 13, 16, 17
Earth–Sun distances of
    perihelion and aphelion for
    35 years 157
Eccentricity of the Moon's orbit
    168–170
Epicyles 1
Espenak, Fred 46
Essen, Louis 3
Euler, Leonard 2
Explanatory supplement to the
    Astronomical Almanac 187
Exposé de l'ensemble des
    théories relatives au
    mouvement de la lune 2
Geocentric ecliptic coordinate
    system 8

Geocentric equatorial
    coordinate system 6, 7
Gravitational force of the sun
    vii, 163
Gutzwiller, Martin Charles 1, 2
Halley, Edmond 1, 44
Hamzaoglu, Melek and Esat
    154
Kepler, Johannes 1, 3, 44
Kepler'second law and
    conservationof angular
    momentum vii
Kollerstrom, Nicholas 168
Laser 3, 6
Law of cosine 7
Lick observatory 3
Lunakhod 1 and 2, 5
Lunar eclipses 44, 45
Lunar Laser Ranging (LLR) 6
Lunar Polynomial Tables 6
McDonald observatory 3, 6
Moon's gravitational force 3
Moon's periods 13
Moonlight sonata, Ludwig van
    Beethoven viii
Murphy, T. W. , Jr 6, 11
Neodymium: Yttrium
    Aluminium Garnet laser
    (Nd:YAG) 6
Neugebauer, Otto 44
New concepts in gravitation vii,
    163

Newton's Forgotten Lunar
    Theory 168
Newton, Isaac
    Principia 1, 119
    The mathematical papers 1,
    168
    Theory of the Moon's
    Motion 1, 168
Obliquity of the ecliptic 3,
    171 - 188
Observatoire de la Côte d'Azur
    (OCA) 6
Ocean tides 3
Orbital motion 2, 153
Orbit of the Moon 147
Peculiarity of the orbit of the
    Moon 3
Perturbation by the sun
    119 - 142
Pic du Midi Observatory 3
Problems with the gravitational
    constant vii
Puzzling motion of the Moon 1
Regression of the nodes
    143 - 155
Retroreflector 5, 6
Roy, Archie Edminston 2, 153
Ruby laser pulses 6
Saros 44 - 118
    length of series 46
    variation of periods 115 - 116
Sidereal month 13, 15

Smith, George Elwood 153
Solar eclipses 44, 45
Speed of light, c 6
Spolter, Pari
   Gravitational Force of the
   Sun vii, 163
   Kep;er's second law and
   conservation of angular
   momentum vii
   New concepts in gravitation
   vii, 163
   Problems with the
   gravitational constant vii
Synodic month 13, 14
The display of all the theories
   relating to the motion of the
   Moon 2
Tides 3
Tisserand, François–Félix 2
Tokyo Astronomical
   Observatory 3

Traité de Mécanique Céleste 2
Transformation equations 8, 9
Treatise on Celestial Mechanics
   2
Tropical month 13
Variation of anomalistic month
   15, 16
Variation of draconic month 16,
   17
Variation of Moon's periods
   13–17
Variation of sidereal month 15
Variation of synodic month 13,
   14
Variation of the advance of the
   perigee and apogee 165
Whiteside, Derek Thomas 1
Wikipedia, the free
   encyclopedia 45

CPSIA information can be obtained at www.ICGtesting.com
Printed in the USA
BVOW11*1856240215

389138BV00001B/1/P